Heinrich Helferich

Atlas of Traumatic Fractures and Luxations with a Brief Treatise

Heinrich Helferich

Atlas of Traumatic Fractures and Luxations with a Brief Treatise

ISBN/EAN: 9783337306731

Printed in Europe, USA, Canada, Australia, Japan

Cover: Foto ©berggeist007 / pixelio.de

More available books at **www.hansebooks.com**

ATLAS

OF

TRAUMATIC

FRACTURES AND LUXATIONS

WITH A BRIEF TREATISE

BY

H. HELFERICH, M.D.,

PROFESSOR AT THE UNIVERSITY OF GREIFSWALD.

WITH ONE HUNDRED AND SIXTY-SIX ILLUSTRA-
TIONS AFTER ORIGINAL DRAWINGS
BY DR. JOS. TRUMPP.

NEW YORK:

WILLIAM WOOD AND COMPANY.

1896.

PREFACE.

THIS Atlas and Treatise are intended to aid students entering upon that important field of surgery which embraces fractures and luxations, and to be a useful book of reference to physicians in their practice. I have endeavored to furnish a work of practical utility, and at the same time to facilitate the comprehension of the questions arising, especially as regards anatomical details.

The first instigation came from the publishers, whose proposition I gladly accepted. On the one hand, I was pleased to utilize in this connection the specimens and drawings which I had accumulated in the course of years; on the other hand, it appeared to me desirable to aid the more general spread of useful knowledge in a department where much harm can be done, and which, at present in particular, is of great importance to every practitioner by reason of novel social arrangements.

I would lay special stress upon the fact that this book is by no means intended to take the place of studies at the clinic or in special courses, but is to form a supplement to the demonstrations and explanations of the instructor.

With very few exceptions, the Atlas presents only original drawings from specimens, many of them

iii

recently prepared. I have striven to utilize the space
at my disposal to the best advantage and to furnish
illustrations both theoretically and practically characteristic and instructive. In these endeavors I have
been well seconded by Dr. J. Trumpp, who undertook
the artistic part of the work and made the original
drawings.

Many specimens illustrating important injuries
were artificially produced, and prepared in the manner I have been accustomed to for years in connection with the operative course on the cadaver. Some
figures show specimens observed by me while assistant to Dr. Thiersch at the Leipsic, Munich, and
Greifswald clinics; others were kindly placed at my
disposal by Professor Bollinger and by my colleague,
Professor Grawitz, from the pathologico-anatomical
collections in Munich and Greifswald.

As the explanations printed opposite the plates did
not appear to me sufficient, the Treatise was prepared, which is printed in separate divisions to ac-
company each section of the Atlas. The lesions of
frequent occurrence and of practical importance are
treated in detail; the rare injuries are explained very
briefly.

It is hoped that the book will be of some use.

H. Helferich, M.D.

Greifswald, October, 1894.

CONTENTS.

LIST OF ILLUSTRATIONS.

I.

FRACTURES.

GENERAL REMARKS ON THEIR CAUSATION, SYMPTOMS
(DISPLACEMENT), AND TREATMENT.

Explanation of Plate 1.

INFRACTIONS (GREEN-STICK FRACTURES).

FIG. 1 *a* and *b.*—*Tibia and fibula of the left leg of a boy* aged 14 (William Kohn), who was severely injured on November 21st, 1889, by being caught between the cam wheels of a threshing-machine. Both bones are drawn as seen from the outside. On the specimen the fracture of the fibula is about three fingers' breadth higher than that of the tibia. At the point of fracture both bones are bent backward so as to produce a salient angle at the anterior side and a depressed angle at the posterior side. Both bones exhibit a marked infraction (green-stick fracture). It may be seen plainly that the bending causes first a separation of the parts on the convex side, and then a detachment of a wedge-shaped fragment on the concave side; this wedge is not fully separated on either bone. (Author's collection.)

FIG. 2.—*Artificially produced infraction of the fibula.* The specimen is taken from the leg of a cadaver, which was fractured by means of Rizzoli's osteoclast. The same effect is produced by other osteoclasts or by breaking a thin bone over the edge of a table. In every instance the base of the wedge, whether completely or incompletely detached, or perhaps sometimes merely marked by fissures, corresponds to the concave side of the bent bone. (Author's collection.)

2

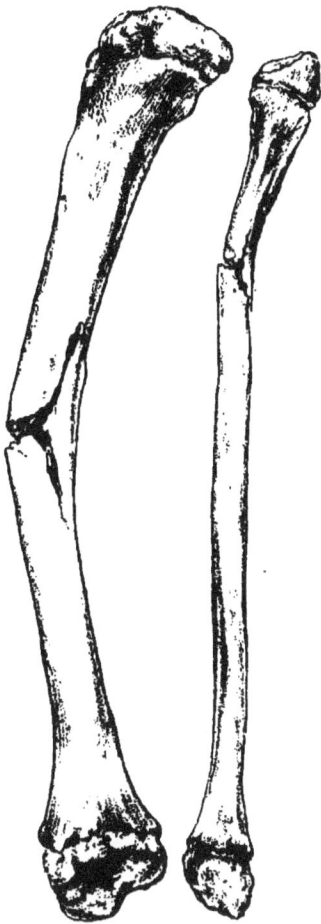

Fig. 1ᵃ   Fig. 1ᵇ   Fig. 2

Lith. Anst. v. F. Reichhold, München

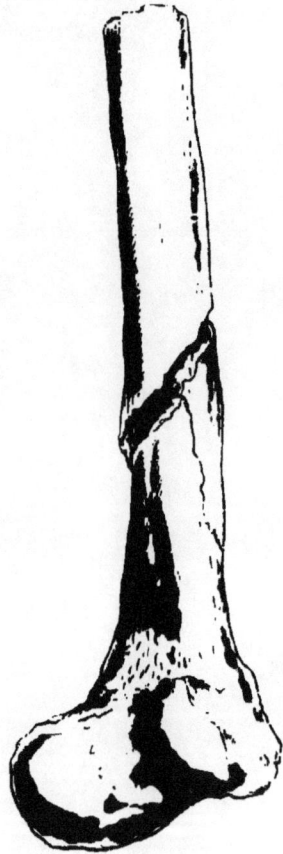

Fig 1
Fig 2

Explanation of Plate 2.

FIG. 1.—*Fracture by torsion of the shaft of the femur in its upper half ;* specimen derived from a woman aged 88 (Anna Kainz). The left femur is seen from in front, and the pronounced spiral direction of the line of fracture is evident. The fracture was caused by rotation of the body while the foot was fixed. Personal observation at the surgical policlinic in Munich. (1884, No. 4,359.)

FIG. 2.—*Artificially produced fracture by torsion of the femur.* The spiral line is seen ascending from below upward to the right; a fissure starts from this spiral, passes almost vertically downward, and its inferior limit again joins the lowest portion of the spiral line. Since near the upper end of the spiral line another vertical fissure passes downward and meets the spiral line again at a lower point, an approximately rhombic fragment is detached, which is a characteristic feature in a large number of fractures by torsion. The short sides of this rhombic fragment are segments of the spiral line of fracture; the long vertical sides comprise about one-fourth of the circumference of the femur.

Fracture by torsion can be artificially produced on the cadaver by thorough fixation of the limb, vigorous torsion, and a sharp blow with a hammer at the point where the fracture is desired. (Author's collection.)

Explanation of Plate 3.

Fig. 1.—*Pronounced fracture by traction. The carpal extremity of the radius and ulna of an adult ;* both styloid processes are torn off in a jagged line. This separation is obviously the result of a sudden traction transmitted through the lateral ligaments, in the present instance caused by injury to the hand in a machine. The separation of the styloid process of the ulna is incomplete. (Personal observation.)

Fig. 2.—*Upper end of the humerus with united fracture by compression.* The head of the humerus and the upper end of the shaft are markedly displaced, but still united by abundant callus. The latter, with its velvety, partly porous structure, can be pretty well recognized in the illustration. The fragment of the head likewise is not normal, but is traversed by fissures at the anatomical neck and within the tubercle, which contribute to the deformity and at the same time exhibit the effect of compression. The upper end of the shaft is displaced forward and inward, and also shifted upward. The fragments are fixed by a spongy mass of callus; the joint was immovable. (Author's collection.)

Fig. 3.—*Comminution of the bones of the forearm at their carpal extremity by powerful machinery.* The patient (Harloff), a man aged 50, was injured on December 21st, 1891, while tending an engine; he stumbled, and his left arm was caught in the drum. As the soft parts were extensively contused the forearm was immediately amputated. The healing of the wound and of a compound fracture of the upper arm which was present at the same time was perfect. (Author's collection.)

4

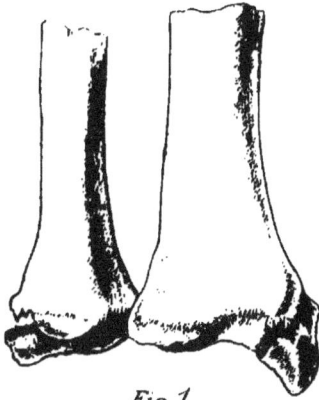

Fig. 1

Fig. 2

Fig 3

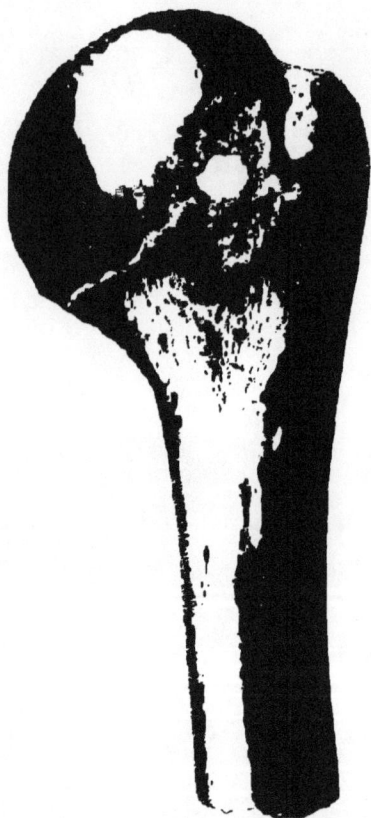

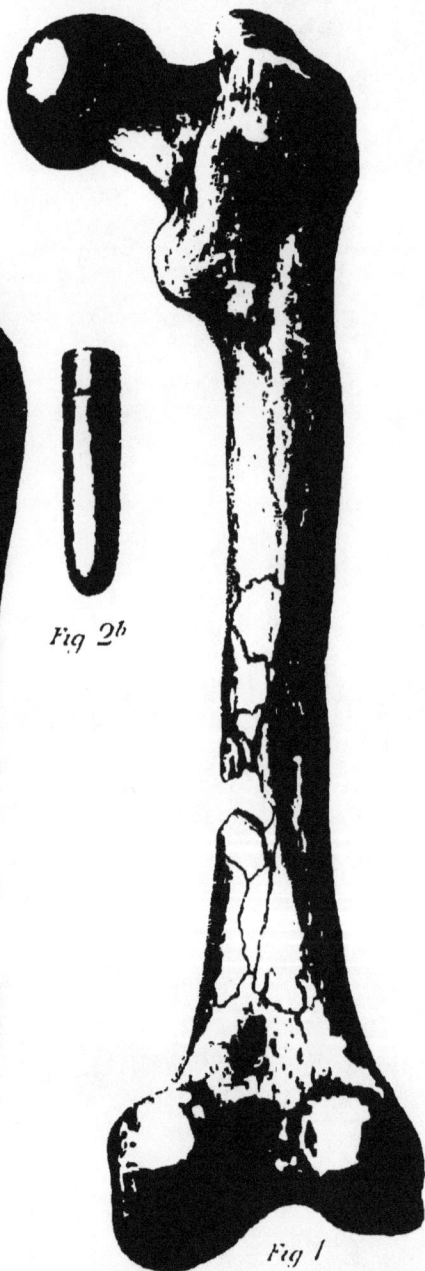

Fig 2^b

Fig 2^a

Fig 1

Lith Anst v F Reichhold, München

Explanation of Plate 4.

FIG. 1.—*Femur with extensive splintered fracture caused by a gunshot wound by the German army rifle, model No. 88, at a distance of 600 metres.* The drawing shows the posterior surface of the femur with the wound of exit and a large number of replaced fragments. At the anterior surface of the femur is the somewhat smaller wound of entrance. Such a comminution of the bone with the associated injury of the soft parts would be an indication for amputation. (Author's collection.)

FIG. 2 *a* and *b.*—*Gunshot perforation of the humerus at its upper end, produced by the army rifle, model No. 88, at a distance of 1,500 metres.*

On the fresh specimen the soft parts, periosteum, and bone showed a smooth perforation. The bullet represented at *b* had entered at the anterior surface of the humerus, made the perforation shown, and was lodged behind under the skin (Fig. 2 *b*).

On the macerated specimen there is a fissure beginning at the point of impact, extending upward and outward through the tubercle, and passing almost completely around the anatomical neck. The wound of exit at the posterior side of the humerus is slightly larger than that of entrance, but still of a rounded form. (Author's collection.)

Explanation of Plate 5.

FIGS. 1 and 2 show different views of the same specimen of a united *fracture of the femur*. It is a good object for demonstrating the forms of displacement, all of which are combined in this one specimen.

The fragments are:

a. Displaced laterally so that their ends do not meet but overlap completely, *i.e.*, *dislocatio ad latus*.

b. These laterally shifted fragments are displaced side by side in a longitudinal direction so that the entire bone is materially shortened, *i.e.*, *dislocatio ad longitudinem cum contractione*.

c. The fragments, however, are not so juxtaposed that their longitudinal axes are parallel, but the axis of one fragment is at an angle to that of the other— the fragments are united at an angle, *i.e.*, *dislocatio ad axin*.

d. Finally, during this multiple displacement, one of the fragments has also rotated on its longitudinal axis. Fig. 1 shows the upper fragment exactly from in front; the lower one therefore is markedly turned inward, *i.e.*, *dislocatio ad peripheriam*.

6

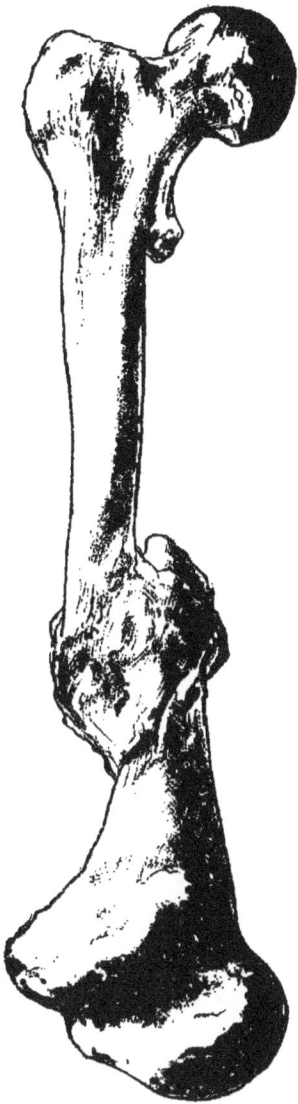 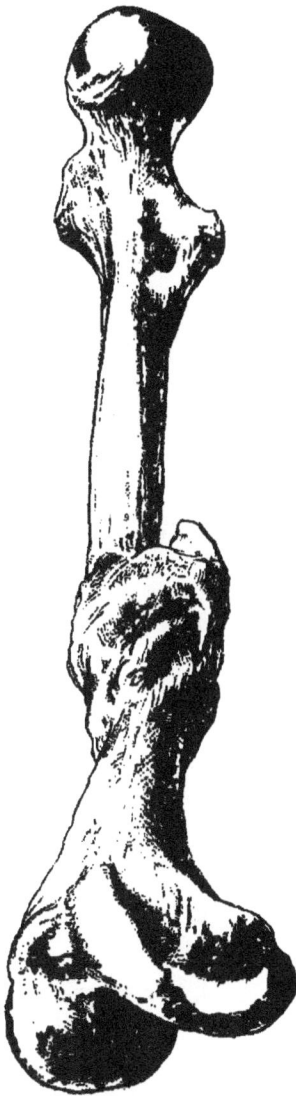

Fig. 1   *Fig* 2

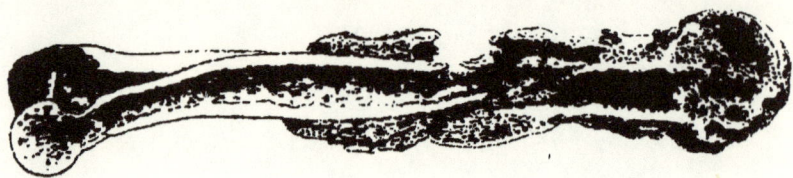

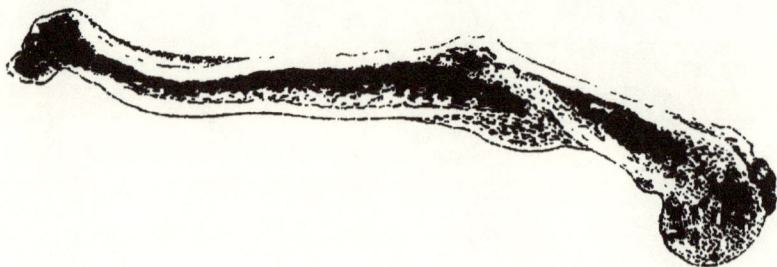

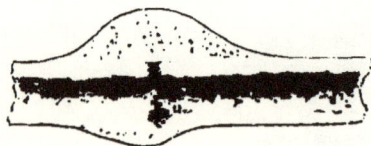

Fig 3

Fig 4

Explanation of Plate 6.

REPARATIVE PROCESS IN FRACTURES, FORMATION OF CALLUS.

FIG. 1.—*Section of a humerus with multiple fractures and extensive callus formation on the shaft.* We recognize in the first place a united fracture by compression at the upper end of the bone, with impaction of the shaft in the cancellated tissue of the head; in the second place, at the inner side of the bone, a transverse fracture of the shaft united to the former by the merged mass of callus. At this point the white compact substance of the bone can be well distinguished from the spongy mass of callus; at the point of fracture the compact bone is thinned by absorptive processes, and at the upper fragment the medullary cavity is shut off by spongy callus; a fragment of the compact bone at the inner side (above) has lost its compact character, looking more like cancellous bone. Although pathological processes took part at the point of fracture, the characteristic relations of the healing of fractures can be very clearly recognized. (Personal observation.)

FIG. 2.—*Section of a humerus with angular union.* We recognize the fractured extremities of the compact outer bone which have undergone spongy alteration, the remnant of callus still occluding the medullary cavity, and the abundant mass of callus in the open angle (concave side) formed by the axial displacement of the fragments. The same specimen is shown (but not in section) in Fig. 3 on Plate 29. (Author's collection.)

FIG. 3.—*Fracture of rib without displacement, with abundant external callus.*

FIG. 4.—*Fracture of the tibia united with displacement.* The callus has become very compact; the medullary cavity has been partly restored. (Figs. 3 and 4 are drawn from specimens in the pathologico-anatomical collection at Greifswald.)

7

General Remarks on Fractures.

IN the discussion of fractures we must distinguish above all those due to an extraneous force (traumatic fractures) from such as occur independently of an extraneous force or from so slight a cause as would not suffice to break a healthy bone (spontaneous fractures).

Spontaneous fractures are the result of fragility of the bone, due as a rule to tumors (sarcomata, metastatic carcinomata, echinococcus cysts, etc.), and to inflammatory diseases of the bone (osteomyelitic necrosis not supported by an appropriate box splint, bone abscess, tuberculous caries, syphilis, rickets, osteomalacia, etc.). The following explanations do not apply to such spontaneous fractures.

The observations here given refer to traumatic fractures of healthy bones.

We distinguish compound fractures and simple or subcutaneous fractures. A compound fracture is one associated with an injury of the skin and soft parts at the point of fracture. As a rule such a complication exposes the seat of the fracture and subjects it to the danger of infection from without; even a slight lesion of the skin and soft parts which does not reach as far as the seat of fracture is included in the term. In these cases the antiseptic or aseptic treatment of the wound must always be carried out strictly in accordance with surgical rules. In this way alone are

1

we justified in expecting a favorable course of such
open fractures which in former times were fraught
with dangers. In other respects the treatment of
these fractures follows the same principles applying
to the simple variety, and aims at a firm knitting of
the broken bone, with the least possible displacement.

It is a matter of frequent experience that this task
is much more difficult in compound fractures, and
that often we have to be satisfied with results which
are not absolutely perfect.

According to the degree of separation of the bone
at the seat of the injury we distinguish complete and
incomplete fractures. The latter include fissures
which traverse the bone without altering its external
form, and infraction or green-stick fracture, which
is observed most frequently on the bent leg bones of
rachitic children, but occurs also in the tubular bones
of adults and in flat bones.

In complete fractures the lines of separation may
pass in very different directions; hence we distinguish transverse, oblique, longitudinal, and spiral
fractures. When small particles are completely detached at the point of fracture, whether or not they
are still connected with the periosteum, the fracture
is called comminuted; but when a larger portion is
broken off at the seat of the injury we may designate
it a separation of a fragment or splinter.

It is a matter of some importance whether the fracture is direct or indirect. This term is used to designate the seat of the fracture with reference to the
force causing the injury. When the fracture occurs
at the point of injury, as for instance in parrying a
blow with the forearm (parrying fracture of the

ulna), it is a direct fracture. But when a fracture of the clavicle occurs in a child from a fall on the hand, it is an indirect fracture. Inasmuch as in a direct fracture the marks of the effective force (contusion and consequent ecchymosis) appear at the seat of the fracture, such injuries as a rule are considered to be more serious than indirect fractures.

A very essential point, morever, is the occurrence of various form of fracture at different ages. It is obvious that the middle adult age furnishes the largest number of fractures, for at this period the heaviest labor is performed and the liability to the dangers and accidents connected with it is greatest. In order to calculate correctly the statistical proportion the number of the population at the various ages must be taken into consideration. We then find that fractures are most frequent between 30 and 40 years (15.4%), and that they are more frequent in advanced age than in childhood; the minimum is found in children up to the age of 10 years. The occurrence of fractures in advanced age is in part the result of an increased fragility due to a senile atrophy of the osseous tissue (diminution of the organic substance in the bone). In early age the presence of the cartilaginous symphysis between diaphyses and epiphyses plays an important part; often enough there is not a true fracture of the tubular bones, but a traumatic separation of the epiphysis, such as occurs spontaneously in inflammatory processes, especially in acute osteomyelitis and also in syphilis.

In considering the mechanism of the production of fractures, the description is to be based upon the study of specimens obtained by accident and those artificially produced on the cadaver. The results thus secured agree with each other; most forms of fracture can be artificially produced without difficulty.

Infraction (green-stick fracture) results from flexion of a bone beyond the limits of its elasticity. In the same way as a stick is broken across the knee and parts first on the convex side, so does a long tubular bone bent in like manner. This happens in various ways. The form of the green-stick fracture is quite characteristic; see Plate 1. An incomplete development of these lines of fracture produces transverse and oblique fractures.

We might perhaps distinguish a special form of infraction caused by lateral pressure upon the end of a bone otherwise fixed; for instance, fracture of the fibula in the typical malleolar fracture by pressure of the astragalus. In part this is certainly a process of flexion.

Fracture by torsion results from twisting. This is possible when one end of the bone is fixed and the rest of that part of the body is twisted. This produces a spiral fracture which can also be effected artificially; see Plate 2. When the bone is twisted to the right the resulting spiral turns toward the right. Torsion causes many oblique and longitudinal fractures.

Fracture by compression is due to a crushing force acting on a bone. This force may be exerted in the longitudinal direction of a tubular bone, in which case there result characteristic infractions at the cancellous end of the bone, as well as complete fracture with impaction of the fragments into each other (for instance, at the upper end of the humerus, tibia, etc., crushing of the calcaneus by a fall upon the feet. See Plate 3, Fig. 2). To the same class belongs also the detachment by contusion of small marginal portions from the articular ends.

Fracture by traction results from the sudden pull of muscles or ligaments in forcible movement of a joint (distorsion). Among characteristic instances are cases of fracture of the patella, olecranon, malleolus, the lower epiphysis of the radius, etc. See Plate 3, Fig. 1.

Gunshot fracture is due to a gunshot injury of a bone. The specimens illustrated on Plate 4 show the extensive splintering in close proximity and the characteristic gunshot perforation at a great distance of the object.

A knowledge of these relations is also practically of great value, occasionally, for the forensic physician. It is obvious that not rarely combinations of the various mechanical effects may be observed in the living patient. Powerful forces (injury by machinery) sometimes produce complete comminution of bones.

SYMPTOMS OF A RECENT FRACTURE.

On inquiring for the symptoms of a fracture pain is mentioned with remarkable frequency—the most unimportant symptom, which moreover would be valueless for differentiating a fracture from a severe contusion were it not that it is sometimes possible to localize this pain at a narrowly circumscribed spot or line of the bone, while in contusion pain on pressure is more frequently felt over a larger surface.

The characteristic feature of a fracture is the solution of continuity of the bone.   This and its mechanical sequelæ form the most important symptoms of a
fracture.

1. The abnormal mobility is the chief symptom,
which is more or less pronounced and marked in most
cases.   It is absent in incomplete fractures, *i.e.*, fissures and infractions, and in impacted fractures.   In
the latter variety the smaller and firmer portion of a
bone is wedged into the cancellous part and so fixed
mechanically that the two pieces again form a single
bone.   This occurs particularly in fractures of the
neck of the femur, though it is met with also at other
articular ends and different tubular bones.   In other
cases, as in fractures of short bones, the ribs, etc.,
abnormal mobility cannot always be demonstrated.

2. Crepitation, the sensation of friction (possibly
also an audible friction sound), on displacing the
broken ends on each other, results from rubbing the
recently fractured surfaces together.   Crepitation is
conditional on the presence of abnormal mobility;
for where the latter is absent, where the fractured
surfaces cannot be displaced on each other, no crepitation can result.   This symptom, therefore, cannot
be perceived in fissures, infractions, and impacted
fractures.   In other cases, in which the abnormal
mobility is not clearly demonstrable, some sort of
crepitation can still at times be noticed with appropriate attempts at displacement.

In other cases, however, the abnormal mobility is
characteristically present, often even very markedly,
and still crepitation is absent.   This is the case when
the fragments are so displaced that they are no longer

in contact (*dislocatio ad longitudinem*), both when
they are separated from each other (diastasis), as
occurs, for instance, with the fragments of the patella, and when they overlap considerably, with pronounced shortening of the entire bone.

Crepitation is absent, moreover, when soft parts
are situated between the movable fractured extremities, that is, when there is an interposition of soft
parts, chiefly portions of fasciæ and muscles. This
ensues when the sharp fractured extremities are
widely displaced and penetrate into the surrounding
soft parts, and during reduction are not completely
freed. The interposed tissue in that case acts as a
cushion which prevents the contact of the fractured
ends.

3. A third very important symptom is deformity
which can generally be seen and felt. This symptom is absent only in the case of fissures and of those
rare complete fractures in which there is no displacement of the broken ends. The deformity is the result
of the displacement of the broken ends. In order to
characterize this, it has long been customary to distinguish different forms of displacement (see Plate 5),
namely :

a. Angular displacement of the fragments (*dislocatio ad axin*) ;

b. Lateral displacement (*dislocatio ad latus*) ;

c. Longitudinal displacement (*dislocatio ad longitudinem*). In the latter case we must distinguish
whether the fragments are drawn apart (diastasis,
dislocatio ad longitudinem cum distractione), as
occurs in fractures of the olecranon and patella, or
whether they override each other with consequent

8 FRACTURES AND LUXATIONS.FRACTURES AND LUXATIONS.

shortening of the entire bone (*dislocatio ad longitu-dinem cum contractione*), as is frequently observed in tubular bones. Diastasis occurs only when the bony framework of the limb is intact, and merely some prominences are subject to a certain muscular trac-tion (patella, olecranon, trochanter, etc.);

d. Displacement by torsion of the fragments or fragment around its longitudinal axis (*dislocatio ad peripheriam*), slight degrees of which are not rarely seen. It occurs in a marked form in fractures of the neck of the femur, and in fractures of the shaft of the femur and radius, in which the peripheral portion of the bone undergoes such displacement when the patient is put to bed.

Further symptoms of a recent fracture are effusion of blood at the point of injury, the above-mentioned pain, and disturbance of function. The latter two are subjective symptoms dependent upon the individ-uality of the patient, and therefore are not decisive.

THE EXAMINATION OF A FRACTURE

should be gentle and rapid. Frequently inspection will establish the fact, so that manual examination of the fragments is required only to settle certain questions. In every case the examination should clearly determine the nature of the fracture, the form and position of the fragments. To this end anæsthesia is often necessary, especially in fractures involving a joint. Whoever in doubtful cases resorts to anæs-thesia (chloroform, ether, ethyl bromide) by prefer-ence, of course with all due caution, will have no cause to regret it: the more accurate and correct appreciation of the conditions will result in a shorter duration of treatment, and besides an exact reduction can be effected at the same time.

An important auxiliary in the examination is mensuration. Since the broken bones are nearly always shortened, the demonstration of a difference in length is of value. This is not meant to imply that a tape measure should be at once applied; on the contrary, the correct way is to make a careful inspection from a proper distance of the injured limb in comparison with the sound one, both being in symmetrical position. After judicious exercise in the clinic and later in practice slight differences can often be better appreciated with the eye than with the tape measure. Still mensuration should likewise be practised.

COURSE AND REPARATIVE PROCESS OF FRACTURES.

A fracture is followed by a swelling of the surrounding soft parts, which is due partly to the effused blood, partly to the infiltration of the tissues. The swelling is greater in proportion to the severity of the injury and to the length of time elapsed between the latter and the replacement of the fragments and suitable position of the limb.

These conditions of course are not without influence upon the system in general. At the seat of fracture are comminuted bone marrow and other tissue elements, together with the effused blood. This is the reason that rise of temperature, *i.e.*, fever, occurs soon after the injury in healthy persons with recent subcutaneous fractures. This fact may be explained by the absorption of small necrosed tissue elements at the seat of fracture, but might be more correctly ascribed to the action of the blood ferment which is absorbed from the extravasation. That fever results from the absorption of blood ferment has been established by experiment.

Smaller or larger amounts of fat enter the circulation from the crushed bone marrow (for fat embolism *vide infra*), which is partly excreted by the kidneys. Therefore in some cases of fracture fat is found in the urine, sometimes associated with albumin and casts.

At the seat of the injury the tumor caused by the effused blood and a kind of inflammatory swelling (œdema) persists for some days; but under correct treatment it subsides markedly as a rule by the end of the first week. The effusion of blood then manifests itself on the skin by its well-known color changes and the tension diminishes. When the swelling is very great the skin at the seat of the fracture is sometimes raised in serous blisters; these do not disturb the normal course when the treatment is correct and no additional complications occur, but they call for careful disinfection of the skin and an aseptic dressing.

At the point of fracture there is formed, or discovered after the subsidence of the swelling, a rounded fusiform tumor which at an early period is of cartilaginous hardness, the ends of which gradually merge into the normal outlines of the bone. This is the so-called callus. As the latter becomes firmer the abnormal mobility of the fractured part diminishes. Finally the broken ends are truly fixed by the callus: the fracture is consolidated.

It is a noteworthy fact that this course forms the
rule. Under normal conditions, both in new-born
children and in most advanced age, the fracture is
consolidated by means of callus. The bulk of its
substance is a product of the periosteum. As the
latter is irregularly torn at the point of fracture,
small portions of it being possibly displaced into the
neighborhood of the fracture, a periosteal proliferation occurs at these places, which is of the nature of
a periostitis ossificans. The medulla of the bone at
the same time is not altogether passive; it likewise
exhibits some degree of callus formation (medullary
callus). If we picture to ourselves this callus formation on a fracture without marked displacement of
the fragments, the external or periosteal callus resem-
bles a mass of mortar laid all around the broken
ends, the internal or medullary callus occludes the
medullary cavity at the point of fracture, and the
two masses are united by the so-called intermediary
callus formed sparsely by the bone itself.

When the fragments are considerably displaced the
callus formation of course is much more abundant;
in such cases the broken ends are at times, as it were,
plastered together by a large mass of callus. The
callus is most scanty in the fractures occurring in
children, in which the periosteum has remained intact so that it forms a closed sheath around the fracture and prevents displacement of the fragments.

While callus formation was formerly divided into
temporary and definitive (Dupuytren), nowadays we
use these terms only in so far as after the healing of
a fracture in the ordinary sense further changes take
place for a long time, by which the anatomical rela-

tions of the seat of the fracture acquire a more definitive character. In other words, after a fracture is firmly consolidated the point of the injury does not continue unaltered for quite a long period. The callus, at first plentiful and spongy, becomes sparser and firmer, gradually assuming the character of compact bone. Whatever is not required in a mechanical sense of the mass of callus and the fragments undergoes slow absorption; of these parts only so much remains as the bone needs for its mechanical function. The medullary canal likewise may be restored. These processes of absorption and ossification are effected very slowly. Plate 6 contains illustrations showing the external callus, the occlusion of the medullary cavity by internal callus, also callus tissue of a spongy and compact character, and the absorption of old compact bone substance.

UNTOWARD ACCIDENTS IN FRACTURES.

Mention has been made above of *fat embolism*. While the absorption of small quantities of fat in fractures is very frequent and as a rule harmless, the absorption of larger amounts of fat may be very dangerous and even fatal. The fat is derived from the comminution of the bone marrow, sometimes perhaps also from the damaged panniculus adiposus at the seat of the fracture. The fat, which is liquid at the temperature of the body, may pass directly into the ruptured veins of the bone and thus into the circulation; in part it may also come to be absorbed and carried along by way of the lymph channels. The fat then enters the blood current and leads to fat embolism in the pulmonary capillaries. Whatever fat passes through the pulmonary capillaries enters the arterial circulation, where it may cause embolism in the various organs (general fat embolism). In fatal cases extensive fat embolism has been demonstrated in the lungs, in the central nervous system, or in the capillaries of the major circulation. The treatment should be directed toward strengthening the activity of the heart by stimulants so as to favor the excretion of the fat by the kidneys.

Venous thrombosis and embolism in subcutaneous fractures are rare but grave accidents. Cases have been reported in which, in the course of a healing fracture, death occurred suddenly with symptoms of asphyxia; the autopsy showed embolism of the pulmonary artery due to venous thrombosis in the region of the fracture. Other cases gave rise, in a similar manner, to embolic infarction of the lung, and in some cases which recovered the diagnosis of embolism of the pulmonary artery could also be made from the clinical symptoms. Venous thrombosis in the region of the fracture often causes an œdematous swelling of the injured extremity. This accident has been most frequently observed in fractures of the lower extremity (generally in the third week), at times in relatively mild cases, as for instance recently after fracture of the patella.

Lesions of the blood-vessels are very rare; they may cause profuse effusions of blood, and, when the arteries are involved (rupture of the anterior and posterior tibial arteries have been most frequently observed), aneurisms and gangrene. Gangrene due to too tight bandaging will be discussed hereafter.

Nerve lesions may result in various ways in cases
of fracture: for instance, a nerve trunk, such as the
radial and peroneal which rest upon the bone, may
suffer simultaneous injury by the force which causes
the direct fracture; or a nerve trunk may be wounded
by the displaced fractured ends (interposition); or
else during the healing the nerve is compressed,
sometimes almost surrounded, by the callus formation. The symptoms of course depend upon the
cause and the distribution of the injured nerve.
Operative interference (liberation of the compressed
nerve from the callus mass) is not objectionable and
has repeatedly terminated in complete recovery.

Delayed Callus Formation.—While callus forms
sometimes in excess and, though rarely, produces
true tumors (osteoma, enchondroma), its development is occasionally remarkably retarded. The
cause of this delay can seldom be ascertained. Practically it is important that in such cases careful ex-
pectancy and the employment of appropriate measures
will as a rule result in consolidation. Among these
measures are, besides a suitable strengthening diet,
walking about of the patient and suspension of the
broken limbs in appropriate dressings. A favorable
effect is often produced by establishing venous hyper-
æmia at the seat of the fracture by the application of
a moderately tight rubber tube (drainage tube) above
the fracture, while the distal extremity of the limb is
protected by bandaging. More vigorous measures
are friction of the fragments against each other
under anæsthesia or perhaps the insertion of nails
into them in order to set up an irritation and a
stronger reaction.

Pseudarthrosis is the term applied to the false joint which may result when the fracture does not consolidate. Some remarks on this subject will be found under the head of treatment. Briefly it must be remembered that the formation of a false joint may be due to general or local causes. Chief among the general causes are syphilis, general debility, etc. At the seat of the fracture various factors may give rise to a pseudarthrosis, mainly extensive local contusion, such as occurs in serious direct fractures, especially the compound varieties. When the callus formation is permanently at a minimum the formation of a false joint will be the natural consequence. In other cases callus formation may be normal and even excessive and yet a false joint may result, namely, when soft parts are interposed or when the fragments are so displaced that they no longer come into sufficient contact; therefore this accident is more common with the humerus and femur than in limbs containing two bones. It is readily understood that defective immobilization of the fracture likewise favors the occurrence of a pseudarthrosis.

In the treatment of a false joint the minor measures, such as friction of the fragments, the insertion of nails or ivory pins, are usually insufficient; as a rule resection of the fractured ends, possibly followed by a bone suture, will be required. When there is a marked defect of bone at the seat of the fracture healing can be effected only by transplantation of bone between the fragments.

TREATMENT OF FRACTURES.

The treatment aims at recovery without displacement and with good function, that is, consolidation of the fracture with the fragments in good position, without injury to the adjoining parts, especially the neighboring joints. This aim nearly always requires, besides replacement of the fragments, an appropriate dressing which must put the fracture at rest, and therefore must include not only the broken bone but also the two neighboring joints. The dressings may consist of pillows, box splints, wire cradles, and more complicated apparatus; in case of necessity and for the first transportation the broken arm may be fastened to the thorax, the broken leg to the healthy one. As a rule use is made at present of circular hardening (particularly plaster of Paris) bandages, or of splints, or of extension by weights.

There is no question that fractures may be treated in various ways, by the exclusive use of one or the other method, with excellent results, if the surgeon possesses some skill and experience; but in order to avoid unfortunate sequelæ it is desirable that he proceed in general according to definite principles. In early times physicians sometimes inclosed the recent fracture in plaster of Paris on their first visit and left the dressing for weeks undisturbed until the fracture was supposed to be consolidated; this is wrong in principle, and recovery with more or less marked displacement is the necessary result. The first dressing of a fracture must be based on the fact that the place of the injury is increased in thickness by the swelling of the soft parts, which is sometimes considerable; in order to allow for this swelling, the first dressing must be well padded. Of course it should be correctly applied and include the neighboring joints, but make allowance for the greater volume by loose material, such as wadding or the like.

About the eighth day the first dressing should be changed; for then the swelling has certainly partly subsided, and the dressing, having become loose, is apt to permit displacement of the fragments. The new dressing is applied, after careful correction of the position and with slight padding. For the latter I prefer the wood felt supplied by the firm of Hartmann in Heidenheim, as it is both soft and firm, and keeps the skin dry. This dressing likewise is not to be the final one; after about another week, or say two weeks after the injury, the second dressing must be changed. At this time the swelling has fully subsided and the seat of the fracture, though surrounded with callus, is still movable, so that a final correction of the position can be easily effected. This third dressing may in ordinary cases remain until complete consolidation has occurred. After that a light and removable protective dressing may be worn as long as required in each case; best a light splint or a water-glass and chalk dressing cut open.

The dressing of the recently injured limb should not be a circular plaster bandage, unless special conditions obtain and the dressing can be inspected daily. A splint is much better for the first dressing. Disregard of this rule has caused much mischief.

In some cases an excessively tight plaster-of-Paris dressing applied for the purpose of compressing the

fracture has led to ischæmic paralysis and contracture, to gangrene at the seat of the fracture, or even
to gangrene of the whole limb, and many a physician
has in consequence got himself into trouble by being
held responsible for the injury.

All the cases of ischæmic paralysis and contracture
(Volkmann) which I have seen were due to a plaster-
of-Paris dressing applied to the recent fracture. In
such a case the prolonged restriction of the blood
supply to the muscle causes disintegration of its elements, it loses its elasticity, and becomes fixed in its
contracted position (contracture). The irritability of
the respective nerve is intact; that of the muscle, according to the gravity of the case, is more or less
diminished, and at times absent.

Among the splints for fractures flexible metal
splints, or plaster splints (plaster-of-Paris and tow
splints of Beely) especially prepared for each case
(see Fig. 2, Plate 42), will be found particularly useful. Of the former I prefer the wire splints devised
by Dr. Cramer, of Wiesbaden, or padded strips of
tin of different length, width, and thickness. By
keeping these on hand, padded with wadding and
covered with mull, suitable material is always ready
for fixing a broken limb in any position by means of
two such splints and a few bandages. I know that
many of my pupils have these splints in daily use;
they are also employed at the Munich and Greifswald
policlinics.

Extension dressings for the permanent extension
.by weights is correctly employed not only in fractures of the femur, but also in fractures of the upper
extremity (for instance, of the neck of the humerus,

Fig. 1.—Ischæmic Paralysis and Contracture of the Forearm Muscles in a
Young Man, aged 17, the result of a fracture of the lower end of the
humerus about ten years before.

of the elbow-joint), of the spine, etc.   The technique
for all these dressings, of course, must be acquired by
practice, which is readily afforded in every surgical
clinic.

For the treatment of certain fractures other methods
are also in use nowadays, which have given excellent
results in the hands of some specialists, but it is
doubtful whether the methods are suitable for the
general practitioner.   It is unquestionable that the
principle of the suture of the fragments of a fractured
patella gives superior results in the hands of the surgical specialist; it is admitted that the treatment of
fractures of the lower extremities by ambulatory
dressings is followed by good results; for the treatment of the typical fractures of the lower epiphysis
of the radius it is even recommended to dispense
with every dressing and to place the limb simply in
a mitella; but for general medical practice these and
similar methods are not suitable, in my opinion.

After the consolidation of the fracture great importance attaches to the after-treatment, with a view
to restore the function of the injured extremity.   In
this respect a gratifying change is to be noted in
recent times, much more being done in order to
secure good results.   Even in connection with the
later changes of the dressings we may institute careful massage and passive movements of the joints
which have been included in the dressings and have
become somewhat stiff.   Both of these manipulations
come into the foreground after the consolidation of
the fracture; at the same time warm baths, jet baths,
bandaging, and especially the employment of medico-
mechanical apparatus, are of great value.

Particular care is required in the treatment of fractures involving the joints, *i.e.*, those fractures which implicate the articular process of a bone, and therefore give rise to severe lesions of the joint, whose capsule is filled with effused blood. In such cases the aim of the surgeon, the consolidation of the fracture and the preservation of a movable joint, is most difficult of attainment. The indication in these injuries is to change the dressings frequently, during the first one or two weeks every two or three days, later every day. To favor the absorption of the effused blood, unless it has been removed by aspiration, we must resort to slightly compressive dressings, combined with massage whenever these are changed, and, in addition, passive movements, fixation of the extremity in various positions, early active movements, and the application of mechanical apparatus. To carry out such a treatment imposes much labor upon the surgeon, but the reward is a brilliant result when consolidation of the fracture with good mobility is secured.

It might almost appear surprising that I finally discuss badly or, rather, unfavorably united fractures. In spite of every care it may happen to every surgeon that the result of his treatment will be unsatisfactory; besides, the stupidity and intractability of the patients, or their treatment by quacks, furnish opportunities often enough for treating fractures united with deformity. In all such cases improvement of the position should be attempted and forced without loss of time. This will require a refracture of the bone, perhaps by the aid of an osteoclast, followed by an improvement of the position by temporary

manual or permanent extension by heavy weights
and pulleys, and finally the preservation of the favorable position during the renewed consolidation.
Such operative interference is urgently indicated in
badly united articular fractures likewise.

GENERAL REMARKS ON LUXATIONS.

The normal mobility of joints has a limit of excursion which in many cases is not absolute. Every
joint is provided with some arrangement which
checks the continuance of the motion beyond a certain point. This check is effected in some joints by
the form of the bone, in others by articular ligaments, and in a few by the muscles; accordingly we
use the terms muscular, ligamentous, and bony
checks of articular mobility. While the bony check
is absolute, the muscular check varies with the elasticity and distensibility of the respective muscles.
We need but recall the great mobility of the wrist-
joint, for instance, in professional piano players and
the movements of the so-called India-rubber men;
such mobility can be attained only by practice and
the lessening of the muscular check.

Every joint has its limit of mobility, and when the
motion is continued beyond this the articular apparatus suffers an injury, laceration of portions of the
capsule or the ligaments, that is, a strain or sprain
(distorsio). When this lesion of the articular apparatus is extensive it may result in a dislocation
(luxatio), in which the articular extremity of one
bone entirely severs its normal contact with the other
and (with few exceptions) passes more or less com-

pletely (luxatio, subluxatio) through the ruptured capsule.

As in the case of fractures so in luxations we distinguish traumatic, pathological or so-called spontaneous, and congenital forms. The latter are due to true faults of development or to displacements which occurred in utero. Spontaneous luxations result only in severe alterations of the joints by pathological processes, especially by tuberculous caries or extreme stretching of the capsule and ligaments.

Traumatic luxations, which are the only ones to be considered here, result from injuries affecting the joint directly or indirectly; some luxations even are due to active muscular action in sudden violent movements.

Luxations are naturally more frequent in men than in women, and in adults, to the onset of senile age, than in children. In children under ten years luxations are extremely rare. It is noteworthy, too, that, according to Krönlein, among 100 luxations 92.2 affect the upper extremity, 5 the lower extremity, and 2.8 the trunk.

Luxations by the direct action of a force are rare. In such cases the trauma acts upon the region of the joint, where it produces the luxation, as a fracture results in the bone from a direct force. In the occurrence of indirect luxations there is an increase of the joint motion beyond the extreme limit of its physiological excursion, the action of the long lever of the shaft of the bone overcoming the normal check. The short lever (the condyle or the articular extremity which is luxated) then is crowded outward in a definite direction, at the same time forming a fulcrum (the margin of the socket, capsule, ligament, or a neighboring bony projection), loses its contact with the opposite articular surface, and the luxation is accomplished.

We always speak of a luxation of the peripheral portion of the skeleton, for instance, of a luxation of the humerus when the dislocation is at the shoulder-joint, and designate its direction by the course taken by the peripheral bone, for instance, præglenoid luxation of the humerus when the head of that bone has slipped forward in front of the glenoid fossa.

The symptoms of a recent luxation are as a rule very pronounced. The absence of the articular end at its normal position, its presence at an abnormal point, cause at least a very marked deformity, which may be hidden only by a profuse effusion of blood. The position of the dislocated limbs is nearly always quite characteristic, so much so that the diagnosis can frequently be made by simple inspection. In addition the position in the several forms of luxation is as a rule typical, because it is determined by the influence of certain portions of the capsules and liga-ments which are preserved in the regular forms of luxation. The dislocated limb is elastically fixed in this position, that is to say, it may be forced by ex-ternal pressure and traction to the normal limit of its excursion, which has been restricted by the luxation, but when released the limb springs back into the old pathological position.

The last-mentioned symptom is the most important for the differential diagnosis between luxations and fractures, for in the latter this elastic fixation is absent. Other important points in luxations are the absence of the normal bony prominence, the possibility of palpating the articular end in an abnormal position, and the changed direction of the longitudinal axis of the bone. Mensuration is valuable at times, since in some forms of luxation there is no shortening but a lengthening of the limb.

As in fractures so in luxations incidental injuries may be present, such as lesions of nerves and blood-vessels, extensive laceration of the soft parts surrounding the joint, even wounds of the overlying integument which give the luxation an open, compound character. In that event the treatment must be carried out on strictly aseptic principles.

The diagnosis is sometimes rendered very difficult when the dislocation is complicated with a fracture. As a rule this rare complication is due to the fact that the extraneous force continues to act upon the luxated bone, thus causing a fracture of its dislocated end. The treatment of course aims at the reduction of the dislocation. This was formerly done in a very forcible manner by powerful traction with the aid of three or four assistants or the use of block and tackle, which sometimes did much damage (laceration of large vascular and nerve trunks, fractures, etc.); but nowadays reduction is effected in a physiological manner without force, as a rule under anæsthesia. The rule, that the surgeon must effect reduction by making the luxated condyle return by the same way in which it reached its abnormal position, is in the main correct. The manipulations should not be arbitrary, but should be based on an accurate knowledge and observation of the position of the condyle, the rupture of the capsule, and the surrounding soft parts. "The anatomy of the luxation determines pre-eminently our modern procedure" (Krörlein).

While these conditions will be considered at greater length in the special section devoted to this subject, a description of the further procedures after reduction will be appropriate here. Under normal conditions, with a suitable dressing which enforces rest, the laceration of the capsule is repaired, the effused blood is absorbed, and the irritation of the joint (slight synovitis) subsides in a week or two. As soon as possible, even before the end of this period, massage and careful passive movements may and should be begun. If these set up fresh pain and symptoms of articular irritation they may be suspended or continued very gently. Beginning with the third week more extensive movements and active exercises, the use of apparatus, etc., are indicated; finally full restoration of function must be secured.

By habitual luxation we mean the frequent recurrence of the dislocation, often in consequence of the most insignificant injury. Such patients know their condition very well and commonly apply to the surgeon with the correct diagnosis; some of them are able to reduce their luxation themselves. The cause of these habitual luxations is generally a marked lesion of the joint which has left an abnormally widened attachment of the capsule. The treatment recommended is more prolonged immobilization, the injection of alcohol for the purpose of effecting a certain shrinking of the tissues, etc.; in very severe cases resection has been performed. Perhaps arthrotomy and partial extirpation of the capsule might be attempted.

Under certain circumstances a luxation may be irreducible; it may happen that replacement fails in spite of the most careful attempts under anæsthesia. The cause may be the small size of the laceration of the capsule, but usually it depends upon the interposition of adjoining soft parts; that the reduction may be very difficult or impossible when complicated with a fracture of the margin of the socket will be readily understood. In all such cases the luxation should be reduced at an early date by operative interference; the reduction must be forced by opening the joint as far as may be necessary.

When a dislocation has not been reduced the condition presented is that of an old luxation, often enough associated with the formation of a new joint, a nearthrosis. Careful examination and the local condition will decide what steps are to be taken in these cases. When the function of the nearthrosis is quite good, as may happen in rare instances, it may be left undisturbed, and the efforts of the surgeon will be directed toward increasing the mobility of the new joint by appropriate exercises, etc. In other cases the only alternatives are resection or arthrotomy with a view to replace the luxated condyle into the old socket. The latter should be the normal procedure, because such cases of non-reduced luxation will come ever more frequently under treatment, and because the result of reduction is generally far better than that of resection. But it is desirable that reduction be forced as early as possible.

Explanation of Plate 7.

GUNSHOT INJURY OF THE SKULL AT A DISTANCE OF TWO HUNDRED METRES.

FIGS. 1 and 2 give the anterior and posterior view of a skull struck at a distance of 200 metres by a bullet from the new German army rifle, model No. 88. The shot was fired with a full charge of powder at the cadaver.

We see the small, round wound of entrance and the large, jagged opening of the wound of exit; at the latter the mass of the smallest bony fragments could no longer be replaced.

The specimen shows the explosive effect of the modern arms upon bones, especially the skull filled with brain tissue. As is well known, even in the year 1870 the Chassepot rifle at close range produced similar results, so that it was erroneously believed that the French used explosive bullets.

The skull here illustrated is broken into a number of large and small fragments, which are grouped more or less concentrically, separated by many approximately radial lines of fracture, about the wounds of entrance and exit, and carefully reunited by wire.

It is well known that such splintering is now explained by the effects of the hydrostatic pressure, which is manifested particularly upon the skull and less markedly upon tubular bone with large medullary cavity filled with soft marrow. (Author's collection.)

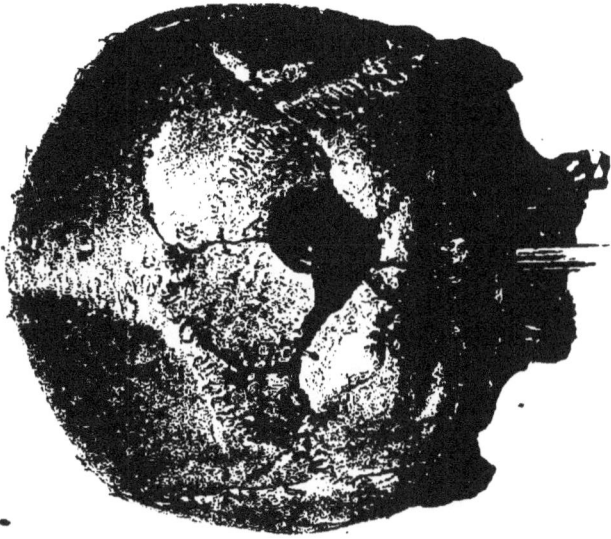

Fig. 1

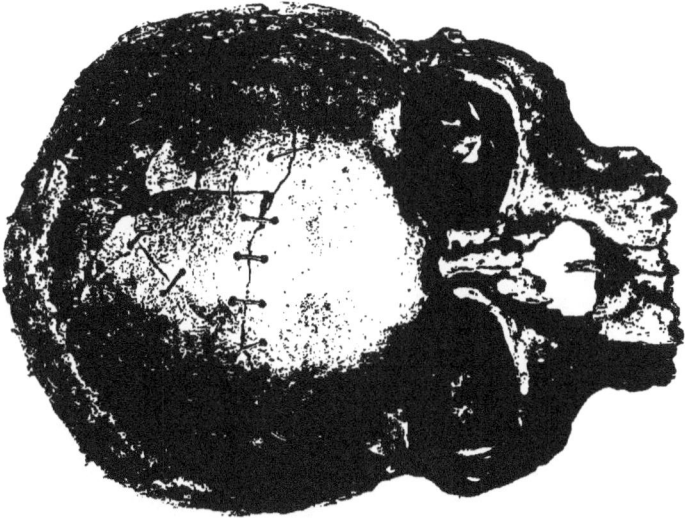

Fig. 2

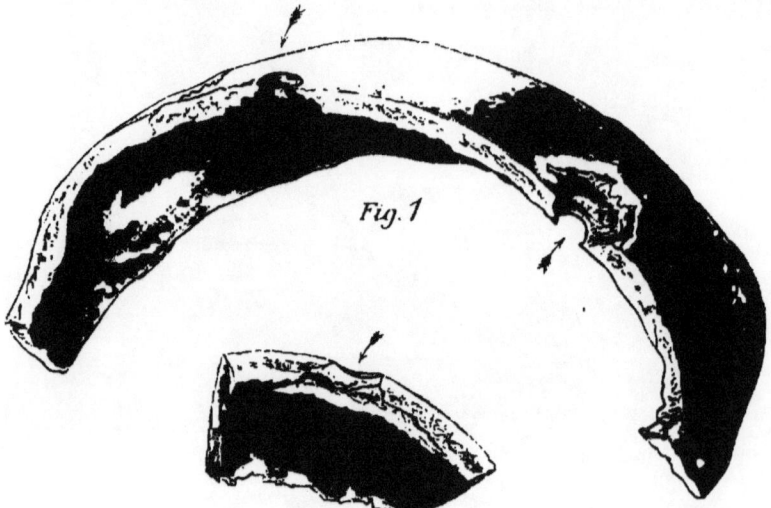

Fig. 1

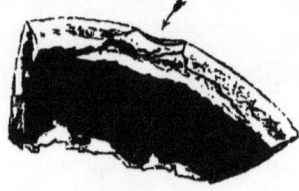

Fig 2

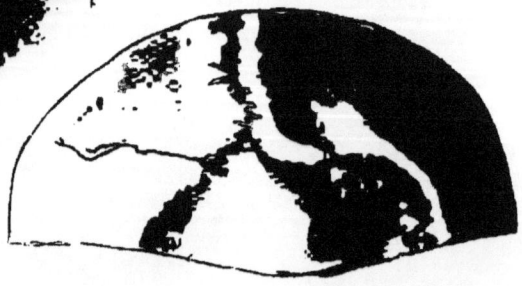

Fig 3

Fig 4

Explanation of Plate 8.

FIG. 1.—*Gunshot wound, from without and from within* (artificial). Portion of the vault of the cranium of a cadaver at which two shots with a small charge of powder were fired, one at the outer, one at the inner surface. Arrows indicate the direction of the bullet. We see that the wound of entrance represents a round perforation, while the wound of exit shows extensive splintering—a larger and irregular loss of substance. The illustration shows at once that the old theory about the brittleness of the glass plate does not apply, and that the result in injuries of the skull rests rather upon purely mechanical conditions. Professor Thiersch, of Leipzig, owns the vault of the cranium of a suicide who killed himself by firing a charge of small shot into his mouth; in that specimen we find the same splintering of the external table as in this artificial preparation. (Author's collection.)

FIG. 2.—*Minor gunshot injury* (artificial), causing merely a slight indentation on the surface of the skull, but an extensive splintering of the inner table. (Author's collection.)

FIG. 3.—Old fracture of the vault of the cranium, united with *depression of the fragments* and thickening of the bone at the point of fracture. We see that the internal table was more extensively splintered than the external. (Pathologico-Anatomical Institute in Greifswald.)

FIG. 4.—Vault of the cranium with *fissure* in the left parietal bone and pronounced *diastasis of the right half of the lambdoidal suture*. The fissure is in direct continuation with the diastasis of the suture. (Pathologico-Anatomical Institute in Greifswald.)

11

Explanation of Plate 9.

FRACTURE OF THE SKULL, WITH RUPTURE OF THE MENINGEAL ARTERY.

FIG. 1.—Section of a skull on which are marked the direction and extent of a fracture observed by myself and immediately transferred at the autopsy by drawing and measurement. The case was that of a laborer (Dittmar), aged 20, who fell from the fourth story on February 7th, 1879, and died of tetanus starting from a severe contused wound in the right trochanteric region. When admitted to the clinic there was a suggillation in the left temporal region, palpable fracture of the squamous portion of the left temporal bone, hemorrhage followed by the escape of cerebro-spinal fluid from the left ear, paresis of the left half of the face and of the right upper and lower extremities. The illustration clearly shows the black line of fracture, adjoining it the grooves for the branches of the middle meningeal artery, and finally the line of suture between the left parietal and the frontal bone; the latter can be best traced from the point where the section passes through the vault of the cranium. Within the line of fracture and the distribution of the posterior branch of the meningeal artery a circular dotted line surrounds the slightly shaded spot which marks the point where the effused blood from the ruptured artery was found between the skull and the dura at the autopsy. (Personal observation.)

FIG. 2.—Horizontal section through the skull with its contents. A large *effusion of blood derived from the middle meningeal artery* is found between the skull and the dura. The figure clearly shows the compression of the brain that may result from such a meningeal effusion. (From Hutchinson's " Illustrations of Clinical Surgery," II., Plate 54.)

12

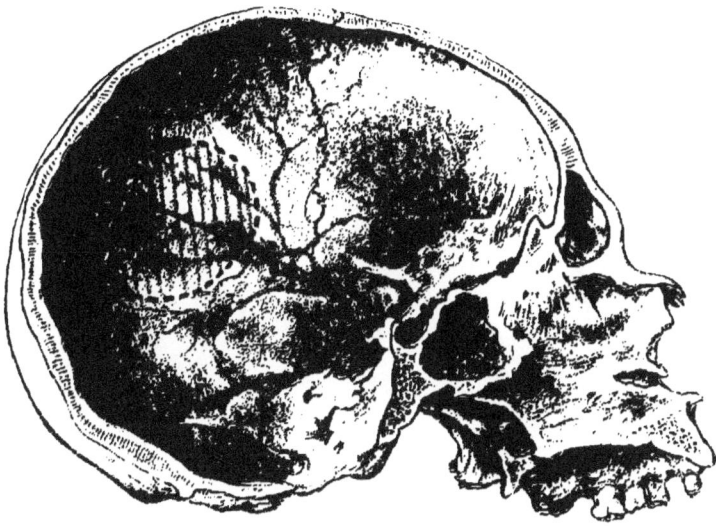

Fig. 1

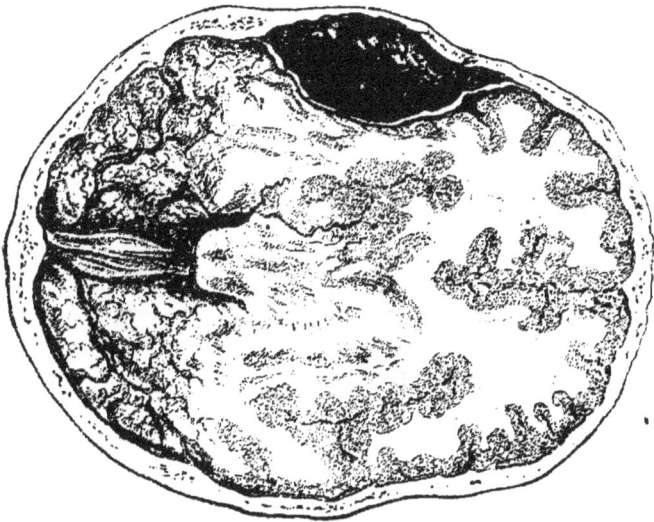

Fig 2

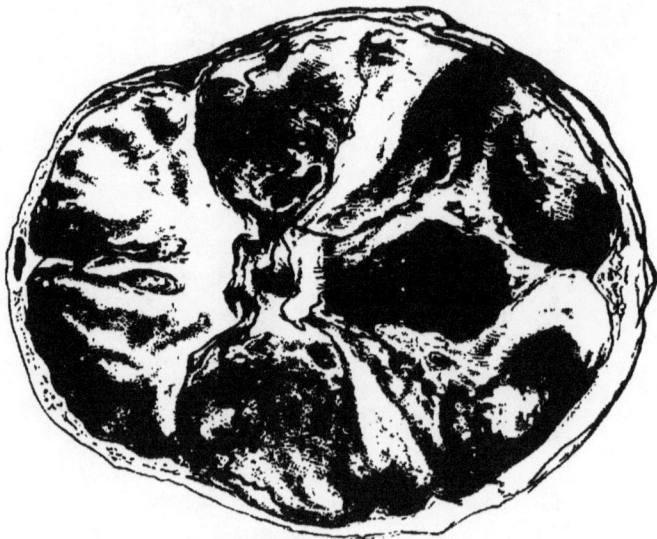

Fig.1

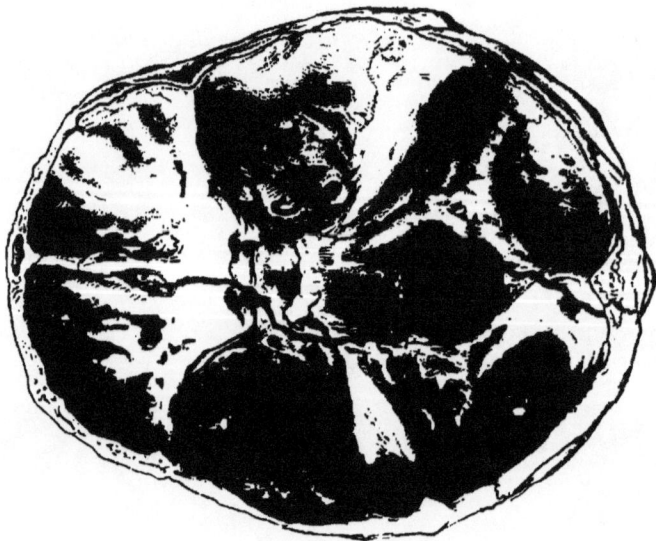

Fig 2

Lith Anst v F Reichhold, München

Explanation of Plate 10.

FIG. 1.—*Transverse fracture of the base of the skull.* This fracture was produced artificially by transverse compression, in a suitable apparatus, of the closed and intact skull of a fresh cadaver. Under these circumstances we can notice, as is well known, first a certain elasticity, a slight change of form resulting, which disappears again with the cessation of the pressure. A moderately increased pressure then causes a transverse fracture of the base of the skull, sometimes passing simply through the middle fossa, sometimes also through the parietal bone, etc. (Personal observation. See also Messerer, "Ueber Elasticität und Festigkeit der menschlichen Knochen," Plate 5.)

FIG. 2.—*Longitudinal fracture by compression of the base of the skull.* The patient, a man aged 35, was injured by a fall upon the head from a height of ten feet. The autopsy showed the fracture of the base of the skull, passing through the foramen magnum, delineated in the illustration. This case reported by Hutchinson corresponds to the longitudinal fractures of the base of the skull which are produced artificially in an analogous or similar form by compression of the closed skull. (Hutchinson, "Illustrations of Clinical Surgery," I., Plate 30.)

13

Explanation of Plate 11.

FIG. 1.—*Sagittal section through the base of the
skull and the left maxillary articulation* (normal).
The illustration shows not only the relations of the
maxillary joint, the articular process of the inferior
maxilla, etc., but also the extremely thin portion of
the base of the skull at this point. It is intended to
show that a force acting upon the inferior maxilla, if
it is transmitted to the ascending portion and especially to the articular processes (as by a fall upon the
chin when the mouth is open), may result in a fracture of the base of the skull, as stated by the authori-
ties. It has even been observed that the articular
process has passed at this point through a broad
fracture into the cavity of the skull. Such injuries
of the base of the skull are not apt to be very frequent,
because the inferior maxilla itself breaks, and besides
the thick margins of the bone serve to protect the
thin part of the bone of the socket. (Author's specimen.)

FIG. 2.—Fracture of the base of the skull by the
pressure of the on-crowding spinal column. The patient, aged 66, had fallen head foremost to the ground
from a considerable height. The on-crowding vertebral column caused the fracture of the base of the
skull around the foramen magnum. This observation by W. Baum (*Archiv für klinische Chirurgie*,
Bd. XIX., S. 381) is, as he himself has shown, of
great importance as illustrating a principle. Similar
indirect fractures of the base can be produced artificially by pressure upon the vertebræ, as I have often
convinced myself.

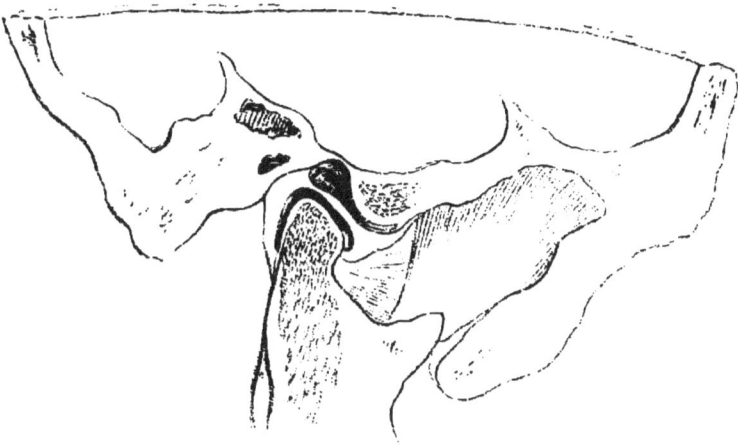

Fig. 1

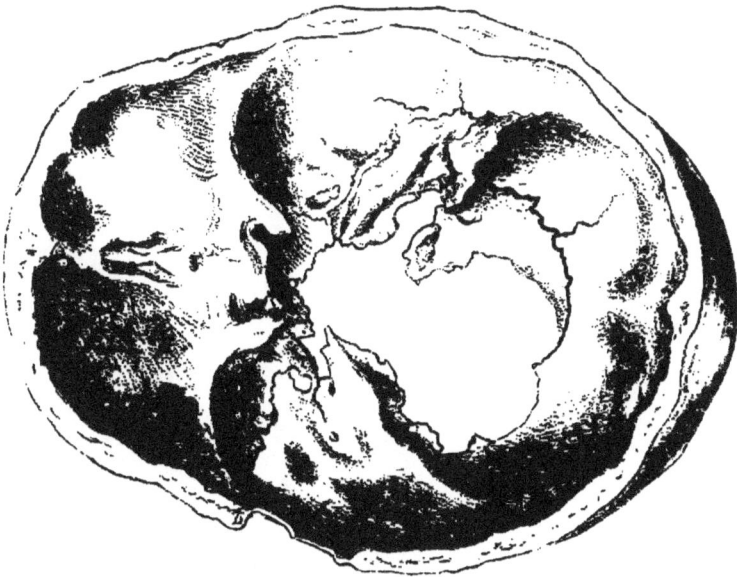

Fig. 2

Lith. Anst .v.F.Reichhold,München

Fig. 1

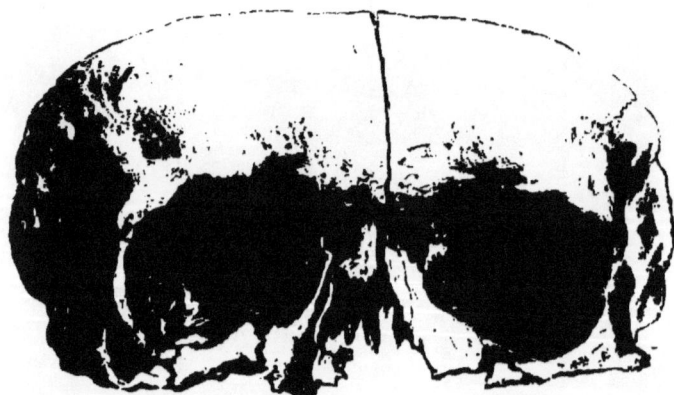

Fig 2

Lith. Anst .v F Reichhold, München

Explanation of Plate 12.

FRACTURE OF THE BASE OF THE SKULL BY INJURY
IN THE NASAL REGION.

FIGS. 1 and 2.—Section and anterior view of a
skull in which a fracture of the base has resulted
from pressure upon the region of the nose and superior maxilla.

The specimen is derived from the cadaver of a man
(Schumann), aged 28, who suffered a fracture of the
nasal bones and the lower orbital margins, and died in
consequence with symptoms of meningitis, on April
12th, 1876. At the autopsy the illustrated remarkable specimen was found, which is now in the collection of the Pathologico-Anatomical Institute in
Leipzig, numbered a 112.

The section clearly shows the upward displacement
of the nasal and cribriform bones, so that the detached
crista galli actually penetrates into the anterior of the
cranial cavity. The anterior view likewise shows
the displacement of the nasal bones, together with
the multiple lines of fracture of both lower orbital
margins.

The illustrations were made from a photograph of
the specimen. (Personal observation.)

II.   Fractures of the Skull.

IT is a noteworthy fact that in fractures of the vault of the cranium the internal plate is always more extensively fractured than the external, and the fragments are more markedly displaced. Formerly this fact was attributed to a greater brittleness of the internal table, for which reason it has also been called tabula vitrea. More recently it has been shown that this phenomenon is based on simple mechanical relations, and that in injuries of the vault of the cranium the plate farthest from the point of impact regularly fractures more extensively. A glance at the illustration on Plate 8 substantiates the important fact that when the vault of the cranium is injured from within, from the cavity of the skull, a like greater splintering as ordinarily occurs on the internal table takes place at the outer table. We must conceive that when a force acts from without, the point of impact at the vault of the cranium bends inward for a certain distance. As soon as the limit of elasticity of this portion of the bone is exceeded, splintering occurs at this point, which we must picture to ourselves as convex toward the cranial cavity, and such splintering must be greater than on the concave side upon which the force acts.

On specimens illustrating fractures of the vault of the cranium we often recognize a position of the fragments which will now be readily understood. The

splinters are depressed below the level of the vault,
and are always more extensive in the depth than on
the surface. On examining severe compound fractures of the vault of the cranium we must expect that
the splintering of the bones in the deeper portions of
the vault, especially about the internal table, will be
far greater than that on the surface. The treatment
of such open fractures of the skull requires in the first
place that the external wound of the soft parts, which
is often very dirty, be rendered perfectly smooth and
clean; this is best effected by careful removal with
knife and scissors of the bruised and soiled portions
of tissue. In the second place the depression of the
fragments must be remedied, and this calls for trephining at the margin of the fracture. In many cases,
in order to secure perfect asepsis, all the splinters of
bone must be extracted. Otherwise the management
of the wound strictly follows surgical rules; the defect at the point of fracture may be covered sooner or
later by an osteoplastic operation utilizing neighboring structures.

The reason why regard to asepsis necessitates such
radical measures and the removal of all splinters of
bone rests upon the possibility or probability that
septic particles may have penetrated from without
between the fragments of the bone; for we find in
some specimens of this character that hairs in larger
or smaller numbers have been imprisoned between
the fragments. I have repeatedly noticed this in
specimens at the pathological institutes of Leipzig
and Munich. The explanation of this fact is obviously that at the moment of the injury the fragments
gape more widely than subsequently, and that the ex-

traneous force at the same instant, after severing the soft parts, presses the hairs into the wound. Thus at the moment when the fragments gape more widely some hairs of the patient may enter between them, and later may be held so firmly as not to be loosened even during the maceration of the bone.

In recent subcutaneous fractures there is hardly ever an indication for operative interference or possibly trephining. Contrary to the views formerly maintained, we now know that minor indentations are not followed under all circumstances by unfavorable results in the brain. The slight diminution of the capacity of the cranial cavity is of no importance. It is true in such cases disturbances may exceptionally occur later on, for instance, the so-called Jacksonian cortical epilepsy, etc., when surgical interference may become necessary.

As regards fractures of the skull in general, it is important that a certain amount of elasticity of the skull, which was demonstrated some time ago by Bruns, has been confirmed by modern investigations, made with every precaution, by means of the best instruments. A force acting upon the skull will produce fracture only when the limit of its elasticity is exceeded. This applies also to the fractures of the base, although it must be admitted that the base is the weakest part of the entire skull.

It is readily understood that fractures of the base of the skull can be produced only in an indirect way. Formerly the theory of contrecoup was advanced in explanation of such injuries. By contrecoup was meant that the mechanical influence of an extraneous force acting upon the vault of the skull caused a certain, somewhat wavelike motion of the surrounding bony parts, and that the continuation of this impulse finally produced the main effect, the fracture, at the opposite point, that is, the base of the skull. This theory of the contrecoup has lost more and more of its importance with advancing knowledge, and nowadays the term contrecoup is hardly used in the sense stated.

Aside from the fact mentioned above, that the base is the weakest part of the entire skull, several other factors are of importance as explaining sufficiently by the mechanical effects the occurrence of indirect fractures of the base. A considerable portion of such fractures result from the extension of fissures from some part of the vault of the cranium. It is perfectly natural that injuries of the head, whether resulting from the impact of an external object or from the patient's fall upon the head, usually cause severe lesions and fractures about the upper or lateral portions (temporal region) of the cranial vault. Every one who has had opportunities for observing a number of similar cases knows that a large percentage of basal fractures are due to this cause; they are simply continuations of a fracture at the vault. Another proportion of basal fractures are caused indirectly by the forcible penetration of parts of the facial bones or of the vertebral column into the base of the skull. If a person has fallen head foremost to the ground and has suffered no direct injury of the cranial vault, pressure upon the base of the skull about the foramen magnum may still have been exerted by the oncrowding spinal column. In this way the skull is fractured as by a direct impact. The same effect will result if the body lands upon the trunk, and the head, as it were, impales itself base foremost upon the vertical spine. Such fractures are extremely characteristic (see Plate 11) and may also be produced experimentally.

Similarly as by the spinal column, a basal fracture may be caused by the facial bones, though such cases are far more rare. Plate 12 shows a specimen in which a force acting upon the nasal region drove the nasal bones into the anterior cranial cavity with characteristic displacement of the crista galli. On Plate 11 the base of the skull near the maxillary articulation is shown in sagittal section in order to recall the fact that this is the thinnest, sometimes translucent part of the base of the skull, where fractures have been observed from pressure of the lower jaw, perhaps due to a fall upon the chin while the mouth was open.

Another group of basal fractures owe their origin to compression of the skull as a whole. It is in these forms that the elasticicity of the skull is manifested. When the compressing force continues to act fractures result, and experiments demonstrate that these fractures assume a longitudinal direction when the compression is longitudinal, and when the latter is transverse the fracture passes transversely through the base of the skull. Of course the lines of fracture are not always identical, but in the main they are similar in character (see Plate 10).

The preceding considerations still leave unexplained the rare isolated fractures of the roof of the orbit and the basal fractures in gunshot wounds. The cause of the latter forms is now admitted to be the effect of hydrostatic pressure. It is not surprising that in the case of such injuries acting upon the skull as a whole the weakest part suffers fracture or fissure. Additional theories respecting these questions will be found in larger works on this subject.

The symptoms of a basal fracture of course vary according to the seat of the fracture or the skull cavity implicated. In the great majority of cases affecting the middle cranial fossa and the region of the ear, there is hemorrhage from the ear of the injured side (laceration of the drumhead), sometimes escape of cerebro-spinal fluid from the ear, trickling of clear serous fluid in considerable quantity after the hemorrhage has ceased (the fluid is free from albumin but contains much chloride of sodium). Frequently we find lesions of the facial and auditory nerves, also of the trigeminus, etc. In fractures involving the anterior cranial fossa the effused blood gravitates in the course of the first few days into the region of the eye and appears as an ecchymosis of the lids; this symptom, however, does not possess the pathognomonic importance formerly ascribed to it. Epistaxis is likewise frequent, and when the patient is resting on the back or when the posterior portion of the nasal cavity is injured, the blood may flow into the pharynx and be swallowed, occasionally causing hæmatemesis. In cases of basal fracture brain symptoms will never be absent. Relatively the slightest affection of this kind is concussion of the brain, a clinical concept, marked by unconsciousness, vomiting, and disturbed cardiac activity, usually slowing of the pulse. The unconsciousness is of variable duration, rarely exceeding twenty-four or thirty-six hours; on regaining consciousness the patient sometimes has forgotten all that has occurred. Otherwise the symptoms disappear completely and recovery follows.

Much more serious is contusion of the brain which implies grave anatomical alterations—hemorrhage into the brain and frequently even laceration of the brain substance. According to the importance of the affected portion of the cortex special nervous symptoms will be present, due to the loss of function of certain centres; then focal symptoms will be associated with the general symptoms. Meningitis and encephalitis are frequent sequelæ.

Occasionally there may also be compression of the brain. Clinical observation and experiments have demonstrated that a relatively large portion of the cranial cavity must be implicated in order to produce symptoms of compression; small extravasations do not cause symptoms of cerebral compression, nor do depressions of the cranial bones unless they are uncommonly large. (Escape of cerebro-spinal fluid.) Such symptoms are caused, in fractures of the skull, mainly by rupture of the middle meningeal artery, in which the extravasation of blood is located between the dura mater and the bone and flattens the convexity of the brain (see Plate 9). In typical cases of this character the early symptoms of compression of the brain subside, the patient regains consciousness, and appears to be on the road to perfect recovery. Then new symptoms are manifested; at first those of irritation, later states of paralysis and depression with renewed loss of consciousness, and finally profound coma. In that event the patient can be saved only by trephining over the seat of the extravasation, removal of the latter, and if necessary ligation of the middle meningeal artery.

In other respects the treatment in uncomplicated fractures of the skull is purely expectant. Rest, good nutrition of the patient, sometimes by means of the stomach tube, perhaps the local application of ice, and especially the prevention of external noxious influences are the sole requirements.

It is still an open question whether disinfecting solutions should be injected in cases of bleeding from the ear. I believe it to be impossible to secure complete disinfection by this means, and I only carefully cleanse the external auditory canal, but thoroughly disinfect the auricle and the surrounding skin, and cover the parts with sterilized cotton. The injection of fluid might be the means of causing infection of the deeper portions of the wound (meningitis).

The union of fractures of the skull is osseous, with remarkably slight callus formation; the latter circumstance is due to the facts that there is little displacement, that the fragments are at perfect rest, and that the dura mater possesses less bone-forming capacity than the periosteum of the tubular bones. Only rarely, in young children, defects of the cranial vault are left behind after fractures; some of them are complicated with meningocele.

III.

FRACTURES AND LUXATIONS

OF THE

INFERIOR MAXILLA, THE THORAX, AND THE VERTEBRAL COLUMN.

Explanation of Plate 13.

FORWARD LUXATION OF THE LOWER MAXILLA.

FIG. 1.—*Bilateral luxation of the lower maxilla*, artificially produced in the cadaver and dissected. The beautiful illustration shows the symptoms of this luxation: the wide open mouth and the chin slightly displaced forward. In addition the specimen shows the position of the articular process of the lower jaw in front of the articular tubercle; the latter projects free, and behind it the socket is empty. Since a portion of the masseter has been removed we see further the joint capsule which passes from the socket to the dislocated condyloid process and is drawn taut. Quite characteristic is the temporal muscle, laid bare by dissection, which is placed in extreme tension by the dislocation and thus leads to an actual incarceration of the articular process in front of the articular tubercle. This illustration elucidates the tension, the impossibility of closing the mouth, and the correct method of reduction (freeing the luxated condyle by downward pressure upon the lower maxilla). (Author's preparation.)

FIGS. 2 and 3 represent *the normal conditions* with the mouth closed (Fig. 2) and open (Fig. 3). The specimen is the same as in Fig. 1, the masseter partly dissected off; the temporal muscle appears more relaxed when the maxilla is simply opened (not luxated), that is to say it is not so tense as in Fig. 1, and still less so in Fig. 2, with the mouth closed.

See also Fig. 1, Plate 11.

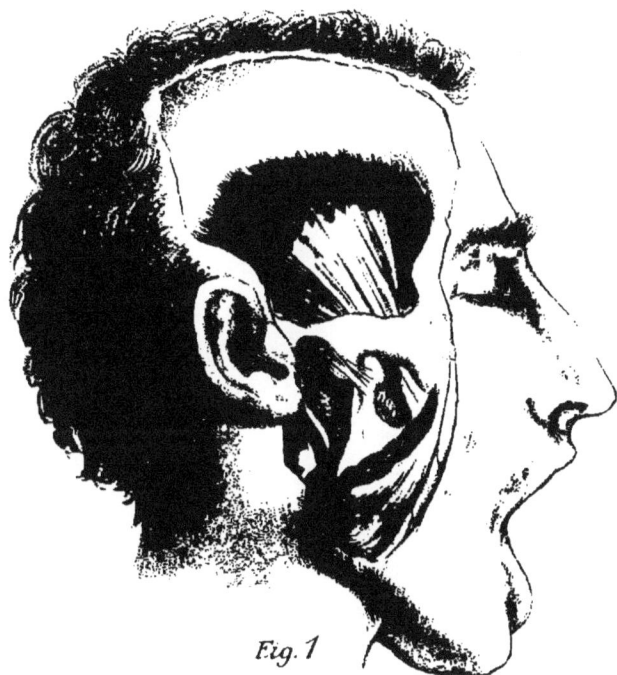

Fig. 1

Fig. 2

Fig. 3

Lith. Anst .v. F. Reichhold, München.

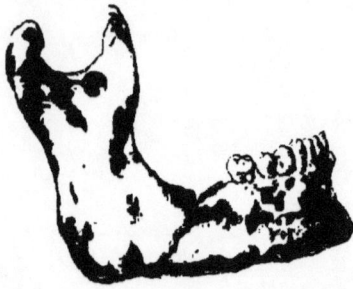

Fig. 1

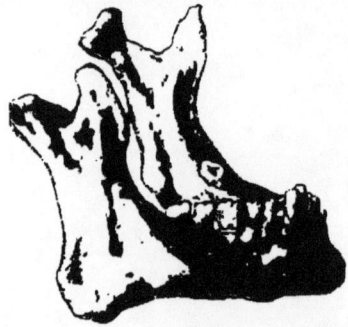

Fig 2ᵃ

Fig 3

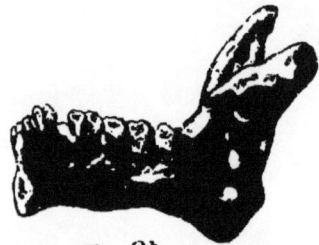

Fig 2ᵇ

Fig 4

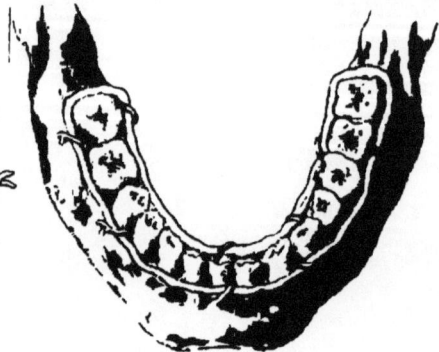

Fig 4ᵃ

Explanation of Plate 14.

FIG. 1.—*Recent fracture in the body of the lower maxilla*, with oblique lines of fracture in the region of the molar teeth, which are missing. (Pathologico-Anatomical Institute in Munich.)

FIG. 2 *a* and *b*.—*Fracture of the articular process of the lower maxilla*. The view of the specimen from within (Fig. 2 *b*) in particular shows the fragment with its pointed end displaced downward, and so united that the upper end of the condyloid process stands below the normal coronoid process. The semilunar fossa is partly filled by the displaced fragment. The influence of these relations upon the position and mobility of the lower jaw can be easily recognized. (Pathologico-Anatomical Institute in Munich.)

FIG. 3. — *Interesting recent oblique fracture through the body of the inferior maxilla and both articular processes*. The latter lines of fracture can have been produced only indirectly by a fall upon the chin. (Compare Fig. 1, Plate 11, with explanation.) The body of the maxilla is fractured at the same time. (Pathologico-Anatomical Institute in Munich.)

FIGS. 4 and 4 *a*.—*Hammond's wire splint for fractures of the inferior maxilla*, shown in Fig. 4 *a* applied in its natural position to the bone. (After Röse, " Ueber Kieferbrüche und Kieferverbände.")

Explanation of Plate 15.

FRACTURES OF THE RIBS AND THE STERNUM.

FIG. 1.—*Four ribs showing old united fractures* on three of them. The fracture is readily recognizable on the upper ribs in the illustration; on the third there was evidently a separation of a splinter which, however, has again united. The fracture is in the region of the angle. (Pathologico-Anatomical Institute in Greifswald.)

FIG. 2.—*Recent fracture of the sternum*, artificially produced in the cadaver and dissected, in analogy with a similar observation by the author. (Personal observation.)

FIG. 3.—*Diastasis between manubrium and body of the sternum united with displacement*, longitudinal section in side view. The specimen is from the collection in the General Hospital of Vienna; it is derived from a woman, aged 42. The displacement of the upper fragment below the lower is easily recognizable. The specimen corresponds exactly with two observations made by me here in living patients; in both cases this fracture with similar displacement had resulted indirectly from a fall on the nape of the neck with forward flexion of the spinal column. The displacement here illustrated could be reduced without difficulty by traction by weights by means of Glisson's suspension apparatus applied to the head, the thorax resting upon a wedge-shaped pillow and the head being slightly bent backward. Union took place in good position. (The illustration is from Gurlt, "Lehre von den Knochenbrüchen," II., S. 273.)

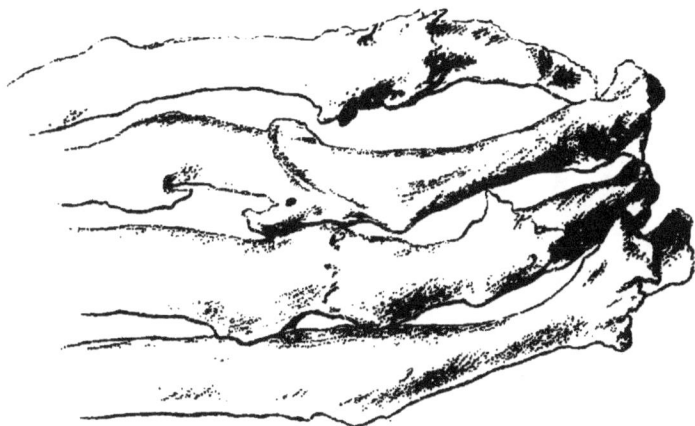

Fig. 1

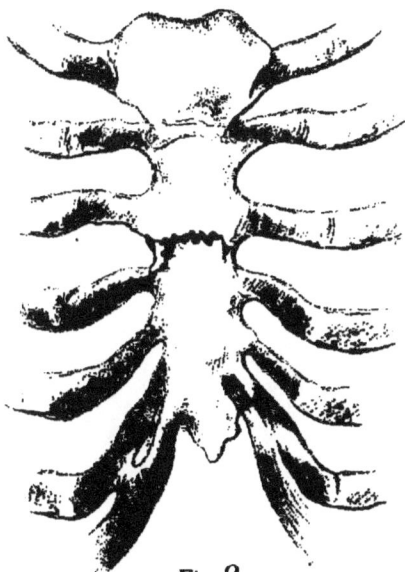

Fig. 2

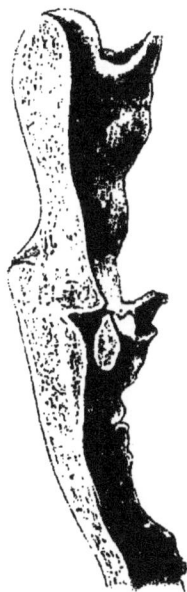

Fig 3

Lith Anst .v. F. Reichhold, München

Fig 1ᵃ

Fig. 1ᵇ

Fig 2ᵃ

Fig 2ᵇ

Lith Anst v F Reichhold, München

Explanation of Plate 16.

LUXATION OF THE CERVICAL VERTEBRÆ.

The illustrations on this plate are drawn from nature. We made a clean ligamentous preparation of the cervical and upper dorsal spine, on which we produced first a unilateral and then a bilateral luxation. Each figure, therefore, is strictly true to nature.

FIG. 1 *a* and *b.—Unilateral luxation (by rotation) of the cervical vertebræ*, lateral and posterior views. It is clearly apparent that the fourth cervical vertebra is so displaced on the fifth that the articular surfaces on the left no longer come in contact. By a movement of abduction (flexion to the right) a diastasis of this joint occurred, and then by a forward rotation a complete dislocation, leading to an interlocking of the two oblique articular processes. The protrusion of the fourth vertebra is evident in the lateral view; the inclination of the spinal column or the head to the right, from the posterior view.

FIG. 2 *a* and *b.—Bilateral luxation (by flexion) of the cervical vertebræ*. Here we easily recognize the marked protrusion of the fourth vertebra beyond the fifth, and the bilateral interlocking, as well as the straight, unchanged direction of the spinal column on the posterior view. (Author's preparations.)

*

Explanation of Plate 17.

Fracture of the Cervical Spine.

This illustration shows a fracture of the cervical spine, involving the sixth and seventh vertebræ, which occurred in a woman, aged 33 (Augusta Ahrens), who was admitted into the Greifswald clinic on June 28th, 1889, and died on July 5th. In accordance with the specimen, which is preserved in the Pathologico-Anatomical Institute, and a photograph taken at the time, a like fracture was produced in a cadaver, from which this drawing was made.

We see clearly the fracture of the sixth and seventh vertebræ, and the pronounced upward dislocation of the seventh vertebra behind, whereby the spinal canal is much narrowed and the cord severely contused.

In the Ahrens case the contusion extended entirely through the cord. Accordingly the symptoms during life were, consciousness undisturbed, sensory and motor paralysis of the trunk and the lower extremities, also disturbances in the upper extremities. The limit of sensibility in front and on both sides was at the height of the third rib. Retention of urine was present. In the region of the fifth cervical vertebra was a distinct backward projection; under anæsthesia this could be easily reduced. The head was suspended by weights by means of Glisson's apparatus and a sling around the head, the patient resting on a portable frame well padded with water cushions. Death occurred with symptoms of paralysis of respiration. (Personal observation.)

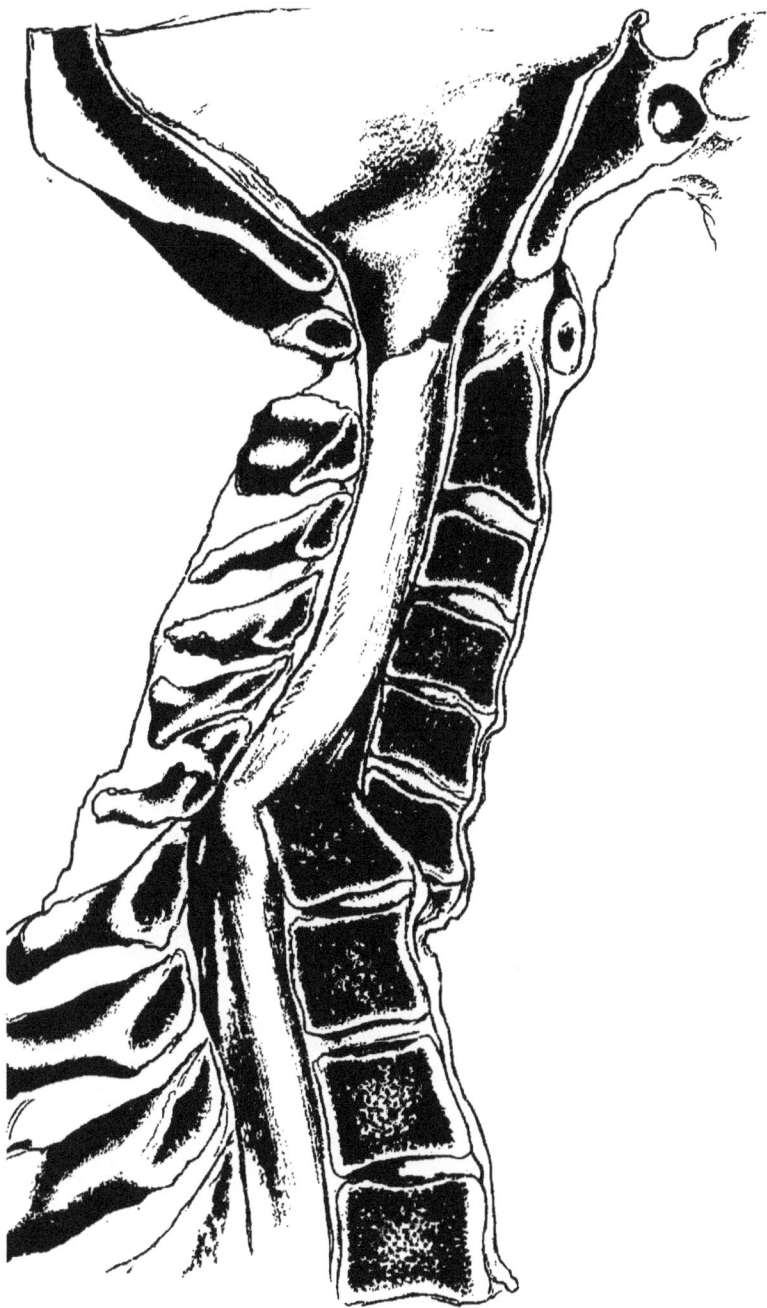

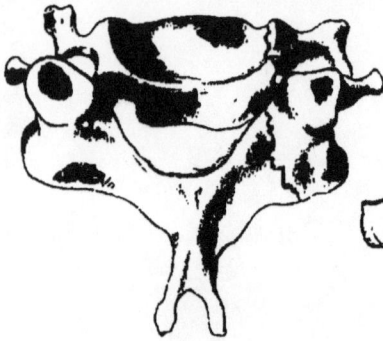

Fig. 1

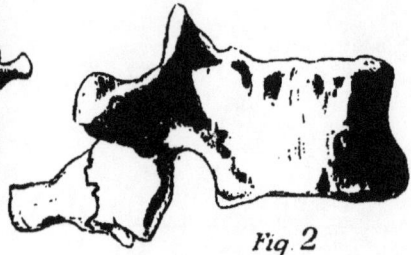

Fig. 2

Fig. 3

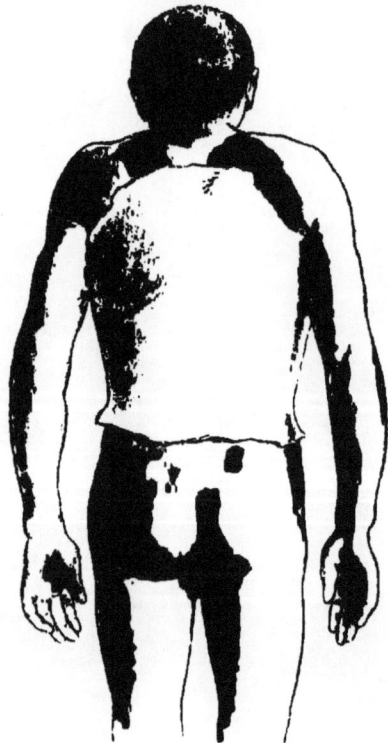

Fig. 4

Lith Anst .v F Reichhold, München

Explanation of Plate 18.

FIG. 1.—*Fracture of the fifth cervical vertebra*, the result of a run-over accident. The body of the vertebra has remained intact. (Pathologico-Anatomical Institute in Greifswald.)

FIG. 2.—*Fracture of a spinous process.* (Personal observation.)

FIG. 3.—*Angular kyphosis by fracture of vertebræ.* G. Wolk, aged 38, on May 24th, 1894, fell from a scaffold five metres high, landing with his back on some bricks. When he regained consciousness he was able, with the support of his comrades, to walk to his home, a short distance from the scene of the accident. When admitted to the clinic, on May 30th, the kyphosis about the eighth and ninth dorsal vertebræ shown in the illustration was found. Palpation in this region caused acute pain. There were no nervous symptoms.

FIG. 4.—The same patient with the plaster jacket applied; this removes some of the weight from the seat of the fracture and protects it against direct or indirect lesions. (From a photograph of the patient.)

III. Fractures of the Bones of the Face.

THESE bones are so accessible to examination from
without or from the nasal and oral cavities that their
fractures hardly ever present any diagnostic difficulties. These fractures should nearly always be looked
upon as compound, since the lesion is in open com-
munication with the nasal or oral cavity; it is noteworthy, however, that the union nevertheless is not
as a rule associated with dangerous accidents.

The nasal bones suffer direct injury only by a blow
or a fall. Fracture of the nasal bones and of portions
of the bony septum lying behind them usually causes
distinct and sometimes great deformity (traumatic
depressed nose). In recent cases the deformity can
be remedied by the insertion of a dressing forceps
into the nasal cavity. Among the symptoms the
suggillations and hemorrhage from the nose are easily
understood; some slight cutaneous emphysema may
result from the passage of air into the cellular tissue
about the seat of the fracture through the laceration
of the mucous membrane.

Fractures of the zygoma and the upper maxilla are
the result of direct lesions, very often from the kick
of a horse's hoof; they are therefore frequently complicated with a wound of the skin. The diagnosis
presents no difficulties; the treatment consists in
keeping the oral cavity clean, careful administration
of a liquid diet, and of course reposition and appro-

37

priate fixation of displaced fragments of the alveolar
process.   This is best effected by the aid of a dentist,
who may often also preserve loose teeth.   Occasionally I have secured good union most simply by nailing a fragment.

Fractures of the lower maxilla are more frequent;
their examination and diagnosis from without and
from the oral cavity are so simple as to hardly call
for remark.   In fractures of the body or the arch of
the lower jaw a typical displacement may be observed
in so far as the posterior portion of the bone is drawn
upward by the action of the masseter, while the anterior portion is displaced downward by the action of
the biventer and the other muscles attached to the
chin.   This is so simple as to require no illustration.
The only difficulty presented by these fractures consists in the retention of the fragments in good position.
Fortunately nowadays we are no longer dependent
upon the dressings, splints, and apparatus applied to
the margin of the lower jaw and the region of the
chin and fastened to the superior maxilla by bandages.   By the aid of a dentist or by simple contriv-
ances the fragments are fixed by small splints fastened
to the teeth of the two broken ends.   Only where the
teeth have been lost or under other special circumstances are we compelled to resort to the older methods or the bone suture with thick silver wire.   Of
course the mouth should be kept as clean as possible.

Among the rarer fractures of the lower jaw those
of the articular process may be mentioned (Plate 14).
The rare fracture of the coronoid process results from
traction of the temporal muscles; union is usually
effected with marked diastasis.

Fractures of the lower maxilla are generally of
direct origin, though indirect fractures may result
from a fall on the chin or from lateral compression of
the bone.

LUXATIONS OF THE LOWER MAXILLA.

Bilateral forward luxation of the lower maxilla is
very frequent.   It results from excessively wide opening of the mouth (yawning, vomiting, etc.).   As is
well known, some displacement of the condyle takes
place with every physiological movement of the lower
jaw; when the mouth is opened the condyle leaves
the socket and reaches the articular tubercle.   The
axis for this movement, that is, the point of least
motion, is situated about at the beginning of the
mandibular canal at the lingula.   When the movement is forced the condyle may pass forward beyond
the articular tubercle, when it again enters a depression in which it is virtually imprisoned: the luxation
is complete.   The powerful traction of the muscles,
especially the temporal, makes the dislocation a very
firm one.

It is obvious from this that the reduction requires
a definite manipulation.   The lower jaw must first
be pressed and pushed downward (best by pressure
with both thumbs inserted into the mouth upon the
alveolar processes of the maxilla) and then forced
slightly backward.   In this way the condyle comes
to rest upon the articular tubercle and the luxation is
reduced.   During the reduction we feel the sudden
cessation of the resistance opposed by the muscles.

The symptoms are exceedingly simple.   The mouth

is wide open, the teeth of the lower jaw project far
beyond those of the upper; the patient is unable to
close the mouth; the prominence of the condyloid
process is absent from its normal position and is felt
farther forward. When the forward luxation is unilateral the mouth is likewise wide open and the chin
is slightly displaced toward the healthy side. The
joint capsule generally remains uninjured and is
merely tensely stretched (Plate 13). This luxation
does not occur in children. The prognosis is favorable, but at times there is a marked tendency to a
recurrence of this dislocation (habitual luxation of
the lower maxilla).

FRACTURES AND LUXATIONS OF THE SPINAL COLUMN.

We may speak of typical fractures of the spinal
column which occur most frequently about the fifth
and sixth cervical and the lowest dorsal and the first
lumbar vertebræ. These fractures are always due to
great violence (fall from a height, imprisonment in a
cave-in, etc.). This is evident even from the fact
that the spine as a whole possesses a high degree of
elasticity and mobility, together with considerable
firmness, for one-fourth of the length of the spinal
column consists of the elastic intervertebral discs,
which permit great mobility. The mobility of the
spine, as is well known, can be materially increased
by practice. We need but recall the extraordinary
movements of the so-called India-rubber men, which
result in almost true flexions in the cervical portion,
at the junction of the dorsal and lumbar portions,

and in the lumbar portion itself.   It is only by great
violence, by displacement beyond the limits of the
possible mobility, with simultaneous muscular fixation of the spine as a whole, that these fractures
occur.

A typical symptom of these vertebral fractures,
aside from some amount of shock which ensues after
such serious injuries, is traumatic kyphosis at the
seat of the fracture.   This results from the displacement of the fragments so as to produce shortening,
which is a consequence of the extraneous force, of the
traction of the powerful longitudinal muscles, and of
secondary movements.   This gives rise to an angular
prominence of the spine on its dorsal aspect, which is
recognized by the characteristic projection of the respective spinous processes.   When the vertebra is
fractured obliquely instead of transversely a lateral
displacement, corresponding to the direction of the
line of fracture, may of course result.

A slight degree of kyphosis is sometimes hard to
determine; in that case the intense local pain is of
importance.   Abnormal mobility and crepitation of
course cannot be demonstrated.   Incidental injuries
may be present in the spinal cord and the nerves
emerging through the intervertebral foramina.
While the spinal cord is well protected in its canal
guarded by bony arches, and by its soft surroundings
in the shape of the spinal dura mater and the cerebro-
spinal fluid, still a more or less grave contusion frequently occurs in fractures of the vertebræ and
displacement of the fragments.   When the contusion
extends through the cord the symptoms correspond
to the distribution of the sensory and motor nerves at

and below the seat of the lesion, and manifest themselves as paralysis of the rectum, bladder, and lower
extremities (paraplegia) in injuries to the dorsal portion; as motor and sensory paralysis of the trunk and
arms, difficult respiration, at times extreme rise of
temperature in lesions of the lower cervical portion;
or by an early fatal termination due to lesion of the
respiratory centre in injuries to the upper cervical
portion of the cord. The prognosis of these fractures
depends upon the nature of the complicating injuries
and their sequels. Fracture of a vertebra *per se* may
heal by osseous union, and many patients recover
from the injury and are capable of more or less hard
work, provided the spinal cord has suffered no damage. But when symptoms of a spinal lesion are present the case is always serious. Even if the patient
escapes a myelitis other dangers threaten: the paralysis of the bladder as a rule requires catheterization
several times a day, and although this should be and
often is done in a truly aseptic manner so that no
harm results, yet in practice it is not uncommon to
have cystitis occur from infection by means of the
catheter; this is followed by the development of a
septic pyelonephritis, caused by micro-organisms, to
which the patient gradually succumbs. Another
danger threatens by way of the anæsthesia of the
paralyzed parts. Bedsores are apt to form, not alone
acutely by the influence of grave trophic disturbances which occur particularly after injuries to the
cervical spine, but also from pressure owing to the
anæsthesia, especially in places where the skin is
often moist, as in the sacral region. No patient requires greater care, more attentive medical super-

vision and watchfulness, than one with paralysis of
a large part of the body due to injury of the spine.
A soft bed free from creases, special protection for
the sacral region, the heels, etc. (water pillows or
millet chaff pillows), frequent change of position by
a partial turn to the right or left side, extreme clean-
liness and dryness of the couch, gentle washing with
alcoholic liquids, sublimate solutions, etc., careful
evacuation of the urine, and watch over the fecal
discharges which the patient passes under his body
(diarrhœa is therefore very unfavorable) are indispensable. Modern hospitals, to which such patients
should always be sent, are provided with special auxiliaries, such as portable bed-frames with an opening
for defecation, and other apparatus for the careful
lifting of the patient (permanent water bed).

The seat of the fracture does not always call for
special care. In fractures involving the cervical
spine useful traction and rest of the injured portion
may be secured by means of Glisson's sling applied
to the head, and extension by weights. The application of a plaster-of-Paris jacket in Sayre's apparatus
has been successfully made in recent fractures, but it
is liable to subject the patient to great risk. Later
on protective apparatus (plaster-of-Paris jackets) are
necessary. Operative interference in order to free
the cord from injurious pressure has been rarely resorted to and is not often indicated.

Other forms of fracture of the vertebræ are of slight
importance. Fractures of the spinous processes
alone, by a direct force, are generally harmless. So
are fractures of the transverse processes; fractures of
the vertebral arches, usually about the lower cervical

vertebræ; fractures by contusion, with spreading of
the body of the vertebra by compression in the direction of the longitudinal axis of the spinal column.

Among *luxations of the spinal column* those in
the region of the dorsal and lumbar vertebræ are extremely rare on account of the anatomical relations.
Luxations of the cervical vertebræ, however, are
more frequent and of practical importance.

Take the cervical vertebræ of a skeleton in their
order and draw through the canal a very thick rubber tube so that the several vertebræ are in contact
with each other. On stretching the tube it will be
easy to separate two of the vertebræ and by appropriate displacement put them in a luxated position.
There is no better way of studying these relations.

We distinguish luxations by flexion and by rotation of the cervical vertebræ (Hueter). The former
result from forced bending of the head against the
chest: in this position the vertebræ spread apart on
their posterior surface, there will be tension and laceration of the ligaments also on the articular processes,
and by a slight simultaneous displacement forward
of the upper vertebra the luxation is effected (Plate
16, Fig. 2). Luxation by rotation is to a certain
extent a unilateral luxation by flexion, yet it does
not result from flexion but from abduction toward
the side remaining intact and from forward rotation
of the upper vertebra (Plate 16, Fig. 1).

The symptoms are at times quite characteristic.
In luxation by flexion the line of the spinous processes is interrupted in a typical manner; sometimes,
it is said, the interval between the vertebræ can be
felt with the finger from the mouth; the neck is in-

variably markedly bent forward and the head is straight. In luxations by rotation the head is always inclined toward the healthy side and slightly turned in the same direction; the displacement of the line of the vertebræ and of the spinous processes is much less pronounced. Injury to the cord is possible in these luxations; as to its results compare the remarks on fractures of the vertebræ. Lesion of the phrenic nerve is absent when the luxation is below the fourth cervical vertebra. The prognosis depends upon the associated injuries and the result of the attempted reduction. In luxations by rotation complicating lesions may be absent.

Treatment.—Reduction is to be effected under profound anæsthesia; in luxation by rotation, by means of abduction toward the healthy side, in order to loosen the interlocking, followed by backward rotation of the cephalic portion on the injured side. In luxation by flexion, first the one and then the other side is treated like a luxation by rotation and reduced. After reduction has been effected, several weeks' fixation by an appropriate dressing is required.

Among other luxations of the cervical spine mention must be made of luxation of the head (luxation between atlas and occiput) by excessive flexion or extension of the head, and luxation of the atlas (between atlas and axis); both of these are generally fatal from complicating lesions.

FRACTURES OF THE RIBS.

Fractures are of course rare in the lowest ribs, which are very movable, and the highest, which

are somewhat protected by their situation; otherwise
they are of frequent occurrence.   In children, owing
to the extreme elasticity of the ribs, fractures are very
rare.

Fractures of the ribs may be direct, or indirect
when the thorax is compressed in the transverse or
sagittal diameter (multiple fractures occur especially
in the axillary line or at the angles).   The diagnosis
is not always possible from the displacement at the
seat of the fracture, but is based rather upon the pain
and a crackling crepitation frequently perceived on
pressure.   Often the lung is injured at the same time.
This organ may be pierced directly by pointed fragments; as this lesion is associated with perforation of
the costal and pulmonary pleura, there is often present
not only hæmothorax and pneumothorax but also a
traumatic cutaneous emphysema which spreads from
the seat of the fracture and in grave cases may dis-
tend the cellular tissue of the whole body.   In that
case the air passes from the alveoli and the smallest
bronchioles of the injured region of the lung into the
pleural cavity during inspiration and expiration and
thence extends farther.   Barring universal cutaneous
emphysema, which may become dangerous by its
great extent, this emphysema is not a serious complication; usually it disappears by absorption in a
few days.   Hæmothorax may require aspiration.

Treatment.—The complications must be attended
to.   Strips of adhesive plaster are to be applied to
the seat of the fracture.   The fracture heals by bony
union, usually without marked displacement.

FRACTURES OF THE STERNUM.

These are either of direct origin, when they are generally very serious owing to the lesion of internal organs, or they result indirectly from forward flexion of the spinal column or of the head so that the chin presses against the upper edge of the sternum. In this way the sternum is compressed in its longitudinal direction and cracked. Fracture of the sternum has also been observed from backward flexion of the trunk, that is, by traction (tearing). This bone being superficially situated, the diagnosis of the fracture is not difficult, especially when the fragments are displaced forward or backward; see Plate 15.

IV.

FRACTURES AND LUXATIONS

OF THE

UPPER EXTREMITY.

Explanation of Plate 19.

SUBCORACOID LUXATION OF THE HUMERUS.

The patient was a man, aged 64, who was injured
about three weeks before. In the mean time the
swelling which must have been present at first has
subsided, and the outlines of the damaged shoulder
can be recognized without difficulty. ` The observer
sitting down facing the patient will see the condition
shown in the illustration and will be able to compare
the details of the diseased right side with the healthy
left side. This comparison is rendered easier by the
fact that a position has been chosen for reproduction
in which the right and left side of the shoulder girdle
are symmetrical; but after studying the details here,
it will not be hard to recognize them also in other
positions of the arm and when the region of the
shoulder is somewhat swelled.

We notice the almost angular projection of the
acromion, and the normal rounded outline of the
shoulder has disappeared. The arm is abducted,
slightly separated from the trunk. The longitudinal
direction of the arm (longitudinal axis) points upward under the coracoid process, or into the region
of the clavicle, instead of to the acromion as on the
healthy side. The external contour of the arm is
somewhat bent in, an angle open toward the outer
side being recognizable. As compared with the
healthy side the arm appears elongated. Finally
under the coracoid process a bulging may be noticed,
which corresponds about to the upper end of the humerus in its changed direction.

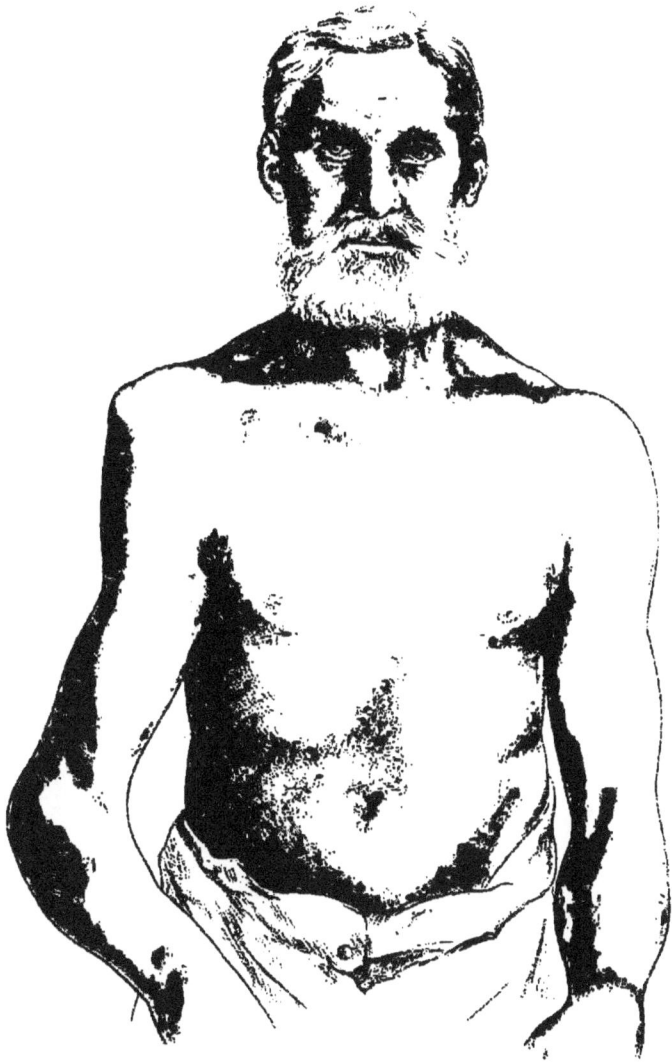

Lith Anst .v. F. Reichhold, München .

Explanation of Plate 20.

SUBCORACOID LUXATION OF THE HUMERUS.

The case illustrated on the preceding plate could
be correctly diagnosticated by inspection, taking
cognizance of the points stated. Palpation confirms
absolutely that the condition is one of subcoracoid
luxation.

Plate 20 shows an anatomical preparation of the
same luxation artificially produced in the cadaver.
The dislocation was effected by extreme abduction,
which resulted first in an axillary luxation and then,
by a secondary displacement, in a subcoracoid luxation. The dissection was made on the dislocated arm.

Here we likewise notice the abduction, the abnormal longitudinal direction of the humerus under the
coracoid process, and its slight bulging at this point.
In addition we may observe with special distinctness
the angular projection of the acromion which is not,
as under normal conditions, overtopped by the rounded
prominence of the head of the humerus. The cause
of the broken outer contour of the arm is now also
apparent: it is the combined result of the direction of
the deltoid muscle, which descends abruptly from the
acromion and is here very tense, and of the external
contour of the lower half of the arm in its abducted
position. In the illustration we can readily distinguish the deltoid muscle, a portion of the pectoralis
major, below its point of insertion the biceps, alongside the latter a portion of the brachialis internus,
and lastly a narrow strip of the triceps.

Explanation of Plate 21.

SUBCORACOID LUXATION OF THE HUMERUS.

This illustration shows a deeper dissection of the preparation figured on the preceding plate. The deltoid muscle is separated from its anterior point of origin and turned over so that the tense portion which springs from the acromion is visible from within. The pectoralis major is likewise detached above and depends loosely between its costal origin and its insertion at the arm; the pectoralis minor lies on its inner surface. The coracoid process is easily recognized; the white coraco-acromial ligament passes outward from this point at the same level; the points of attachment of the coraco-brachialis and the short head of the biceps are distinct. At the lower part of the arm may be seen the stump of the coraco-brachialis from which a piece has been excised, the biceps, and alongside the latter fibres of the brachialis internus. We also see the upper portion of the humerus with the long biceps tendon, and the articular cartilage of the head of the humerus. The caput humeri is partially hidden by the muscles inserted at the greater and lesser tuberosities; the subscapular muscle passes inward and upward, the supra- and infraspinatus outward and upward. Finally the drawing shows the nerve trunks of the axilla, with the axillary nerve passing from behind around the humerus into the deltoid. It will be seen that these nerves are stretched by the subcoracoid luxation and that they are liable to suffer injury from the pressure of the dislocated head. The position of the nerves in the illustration is a product of the dissection; otherwise their course would be slightly different. It is to be particularly recommended that the preparations here shown be reproduced on the cadaver and demonstrated in the various stages, as is done in the author's course of instruction.

28

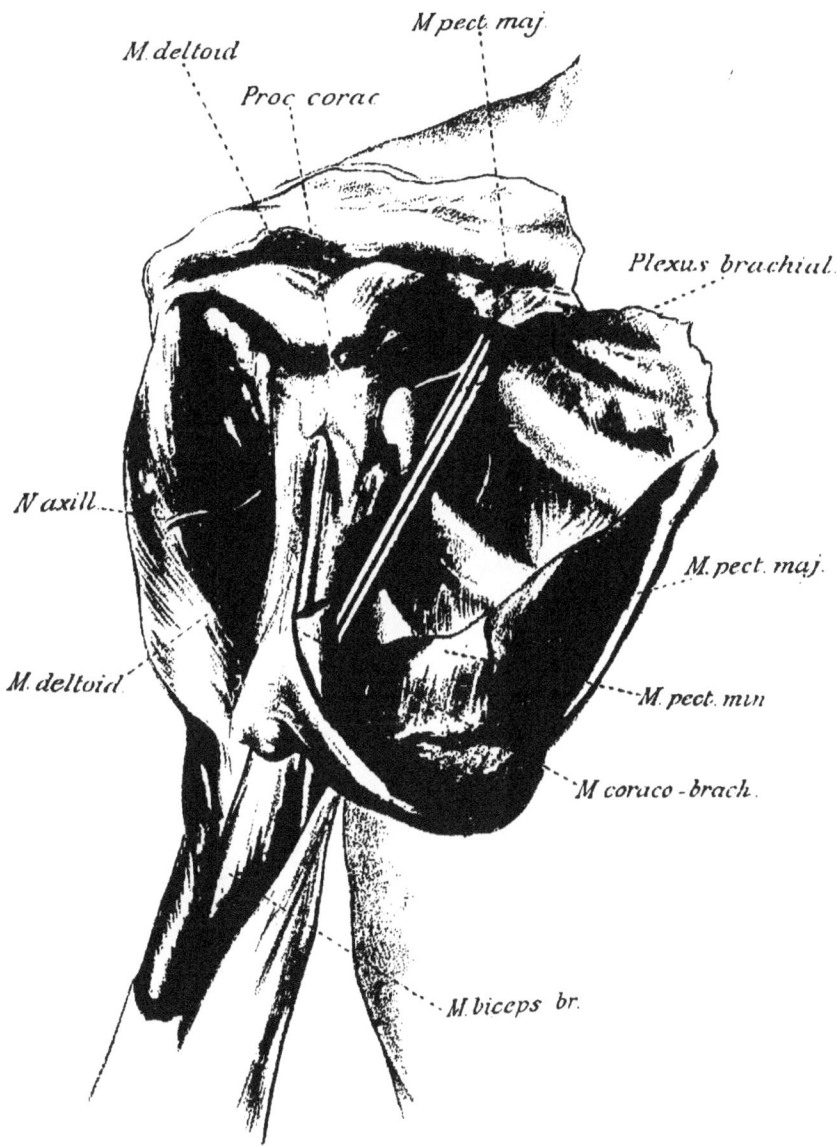

M. deltoid
Proc. corac.
M. pect. maj.
Plexus brachial.
N. axill.
M. pect. maj.
M. deltoid.
M. pect. min.
M. coraco - brach.
M. biceps br.

Lith. Anst. v. F. Reichhold, München.

Fig. 1

Fig. 2

Fig. 3

Fig. 4

Lith. Anst . v F Reichhold, München .

Explanation of Plate 22.

SUBCORACOID LUXATION. OF THE HUMERUS; RE-
DUCTION.

This plate is to illustrate the several steps of *Kocher's method of reduction of a subcoracoid luxation*. This was done with the preparation on Plate
21, each step being immediately photographed. The
drawings on this plate were made from these photo-
graphs.

FIG. 1.—The arm is *adducted* until the elbow
region touches the trunk, which must be straight
(first step). The position of the head of the humerus
has undergone no material change.

FIG. 2.—The arm remaining adducted, it is *rotated
outward* by the aid of the forearm which is bent at a
right angle in the elbow-joint (second step), until the
forearm lies about in the frontal line of the trunk.
Some resistance is experienced and the force employed, of course, must not be excessive. The head
of the humerus during this step is displaced outward,
away from the coracoid process to the acromion, as
is evident in the illustration, especially by the greater
distance of the brachial plexus.

FIG. 3.—The arm, which is kept adducted and rotated outward, is *elevated*, that is, lifted forward
(third step). During this step the head of the humerus begins to pass through the rupture in the capsule and to resume its normal position.

FIG. 4.—In the succeeding *inward rotation* (fourth
step) the head is fully reduced, without the violent
snap which always indicates that the reduction was
forced and not effected in a "physiological" manner.

Explanation of Plate 23.

FIG. 1 shows the two bones in the luxated position, seen from in front. The head of the humerus hides the region of the glenoid fossa and is situated on the anterior surface of the neck of the scapula, below the coracoid process. We see the free anterior surface of the head of the humerus covered with cartilage, and the margin of the osseous proliferation at the neck of the scapula which surrounds the newly formed articular cavity. The humerus is slightly abducted. The mobility of this abnormal connection is extremely small; the cause of this fact becomes clear on closer inspection of the bones at their point of contact.

In Fig. 2 the two bones are seen again, the scapula in anterior view as in Fig. 1, but the humerus turned about 180° so as to present its posterior surface, which faces the scapula. On the scapula the glenoid fossa appears in side view, considerably foreshortened, its anterior circumference diminished by abrasion, and immediately adjoining it the new socket surrounded by the rather uneven bony wall. On the humerus likewise can be observed the depression caused by abrasion against the margin of the glenoid fossa, and at the point corresponding to the anatomical neck some proliferations of bone such as are characteristic of arthritis deformans. The eburnations present at the points of contact of the two bones in the region of the abraded surfaces unfortunately cannot be clearly delineated. (Author's specimen.)

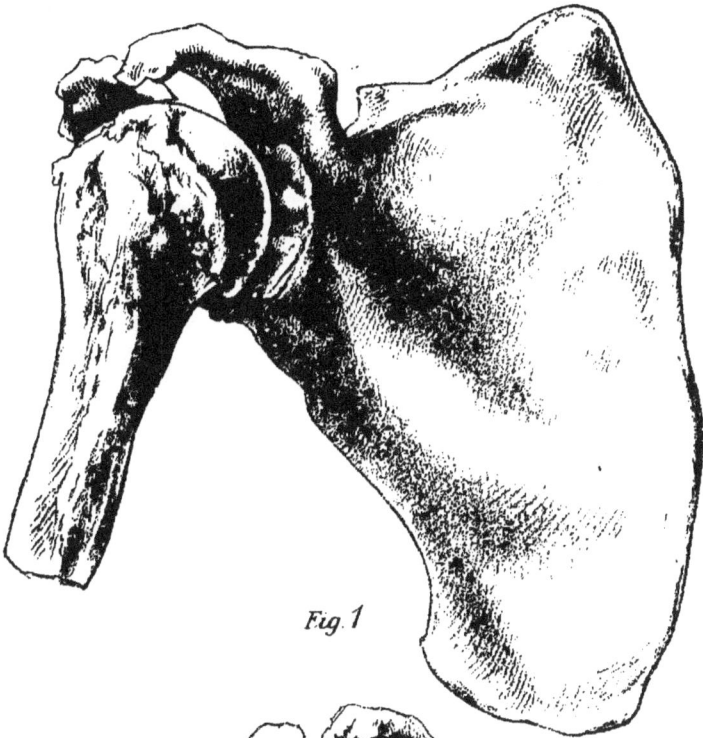

Fig. 1

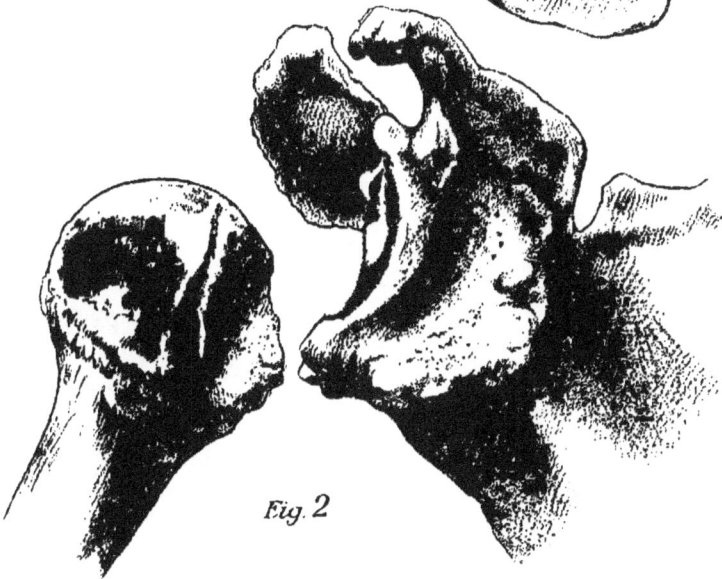

Fig. 2

Lith. Anst .v. F.Reichhold, München .

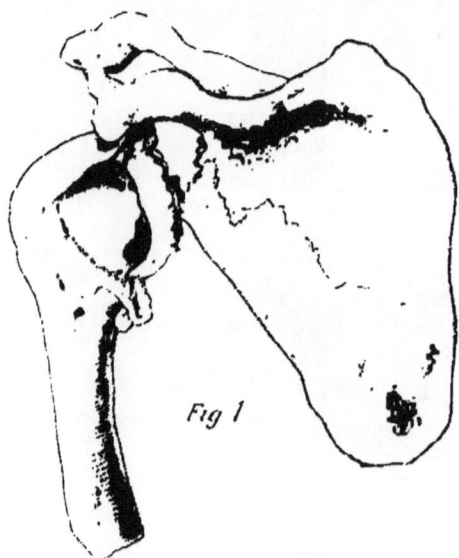

Fig 1

Fig 2

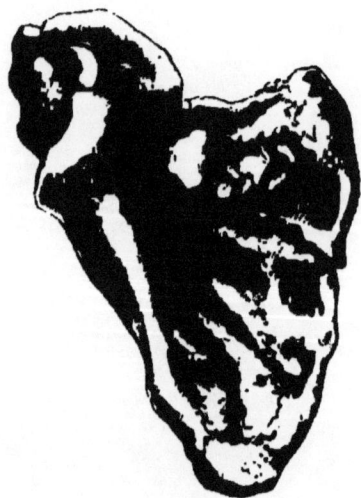

Fig 3

Lith Anst. v. F Reichhold, München

Explanation of Plate 24.

FIG. 1.—*Fracture of the neck of the scapula.*
We see the line of fracture at the neck, the displacement of the small fragment which contains the articular surface, and the natural downward displacement
of the fragment together with the arm. Pushing up
of the arm, however, will suffice to show the mobility
of the fragments and the perceptible crepitation. In
this figure the line of fracture at the neck has been
artificially produced and so drawn; but as a rule the
neck of the scapula breaks so that the coracoid process forms part of the displaced fragment, as appears
in Fig. 2 by the line of fracture passing from the
lower margin of the articular process to the notch.

The other line of fracture shown in Fig. 2 passes
obliquely through the lower margin of the articular
fossa, and therefore indicates the detachment of this
portion from the lower margin of the socket. (Personal observation.)

FIG. 3.—Lines of fracture of the scapula united by
callus; on the posterior surface their course through
the crest of the scapula was very distinct. (Pathologico-Anatomical Institute in Greifswald.)

The study of Fig. 1 on this plate is essential for
appreciating the points which are of importance in
the differential diagnosis of injuries in the region of
the scapula.

Explanation of Plate 25.

LUXATIONS OF THE CLAVICLE.

FIG. 1.—*Upward luxation of the acromial end of the clavicle.*

This illustration is meant to facilitate the diagnosis of this luxation in the living patient; it is mistaken with remarkable frequency for a subcoracoid luxation of the humerus. In our Fig. 1 the right arm is shown slightly drawn up, after a preparation artificially produced in the cadaver so as to correspond with a photograph of a living patient in whom the characteristic displacement was most marked in this position. A glance at the illustration shows the pronounced displacement of the acromial end of the clavicle, which is otherwise intact; the acromion is in normal relation to the arm and the rounding of the shoulder is unchanged in other respects. That portion of the shoulder girdle which is normally kept away by the interposed clavicle from the trunk is approximated to the latter and the axilla is almost obliterated; this approximation of the right arm becomes most marked on comparing its distance from the right nipple, which is much less than on the opposite side.

But when the displacement is less marked and great swelling is present, how is this luxation to be correctly diagnosticated?

A single manipulation suffices: the patient being seated opposite the physician, the latter follows with both hands the crests of the scapulæ on either side from behind and thus reaches the point of the acromion with certainty (Fig. 2). The position of the acromion with reference to the prominence of the clavicle will immediately decide the question.

FIG. 3.—*Forward luxation of the sternal end of the clavicle.* The illustration requires no comment; it is accurately drawn from a preparation on the cadaver.

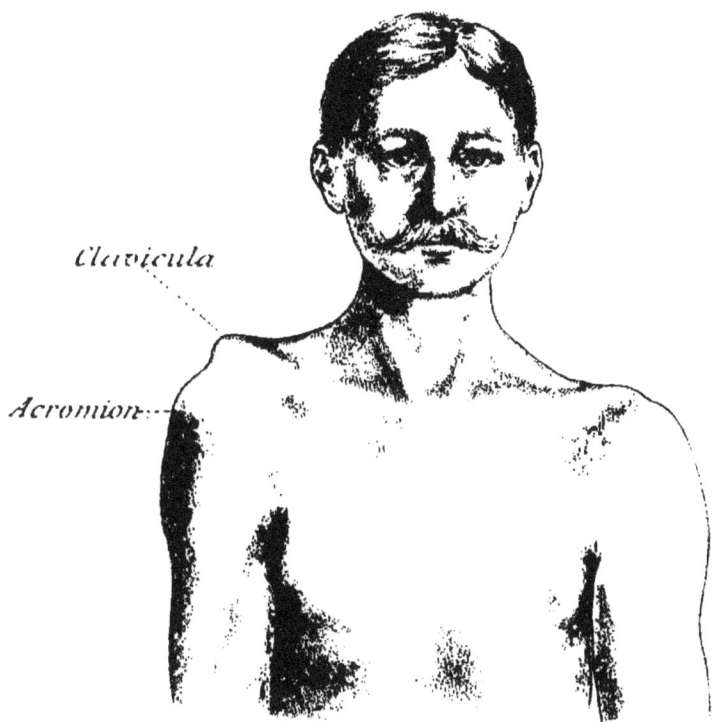

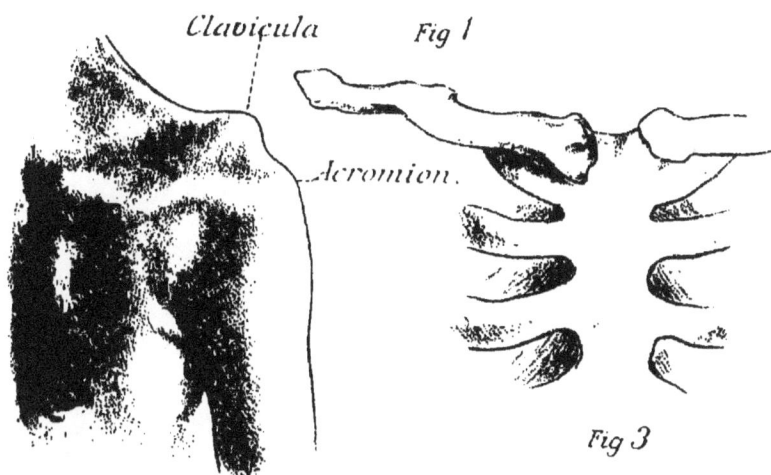

Clavicula

Acromion

Clavicula

Acromion

Fig 1

Fig 2

Fig 3

Lith.Anst.v.F.Reichhold,München

Lith Anst v F Reichhold. Munchen

Explanation of Plate 26.

The seat of the fracture is between the sternal and
the middle third of the left clavicle. The fragments
override, thus shortening the clavicle. The sternal
fragment is displaced upward by the traction of the
sterno-cleido-mastoid; the external fragment is displaced beneath the inner. In the illustration the sterno-cleido-mastoid is easily recognized; postero-ex-
ternally the trapezius limits the outline of the nucha.
The deltoid can be recognized at once; the clavicular
origin of the pectoralis major has been detached and
removed so that in this opening we may see the first
rib, the exact position of the fragments, beneath the
lateral fragment the subclavius muscle, and finally
and particularly the large vessels and nerves. Between the yellowish brachial plexus and the blue subclavian vein a small strip of the subclavian artery
(red) is visible.

As a consequence of the fracture of the clavicle the
position of the arm is changed: it is approximated to
the trunk with obliteration of the axilla and it also
hangs farther down; all this is very evident in the
illustration, but the forward and inward displacement
of the shoulder region cannot be easily represented.

Of especial importance is the depicted relation between the fracture and the large vessels and nerves;
it is readily understood that severe lesions of the vessels and especially of the nerves may result from the
pressure of the fragments.

Explanation of Plate 27.

A glance at Fig. 1 on the following Plate 28 shows the course of the epiphyseal line. Obviously, as in Fig. 1 on this plate, we are dealing with bones of young persons in which the epiphyseal lines are still present; the coracoid process, too, has not yet undergone osseous union. The illustration is very instructive as showing how the epiphysis remains in contact with the scapula by means of the ligamentous apparatus of the shoulder-joint and by the muscles inserted at the tuberosity of the humerus; some shreds of periosteum adhere to the epiphysis. The diaphyseal end in its characteristic form is drawn underneath. Often enough the separation of this cartilaginous union is so extensive as to lead to a marked and serious displacement of the shaft of the humerus, especially forward and inward; sometimes it is remediable in no other way than by operation.

FIG. 2.—Picture of a man (Bertram, 1878) with considerable shortening (arrest of growth) of the right humerus in consequence of an injury at the upper end of the bone in early youth. As is well known, lesion of the epiphyseal cartilage is not rarely followed by an arrest of its physiological production, that is to say, diminished growth of the injured bone; this is most marked when the union took place with the fragments much displaced. (Personal observation.)

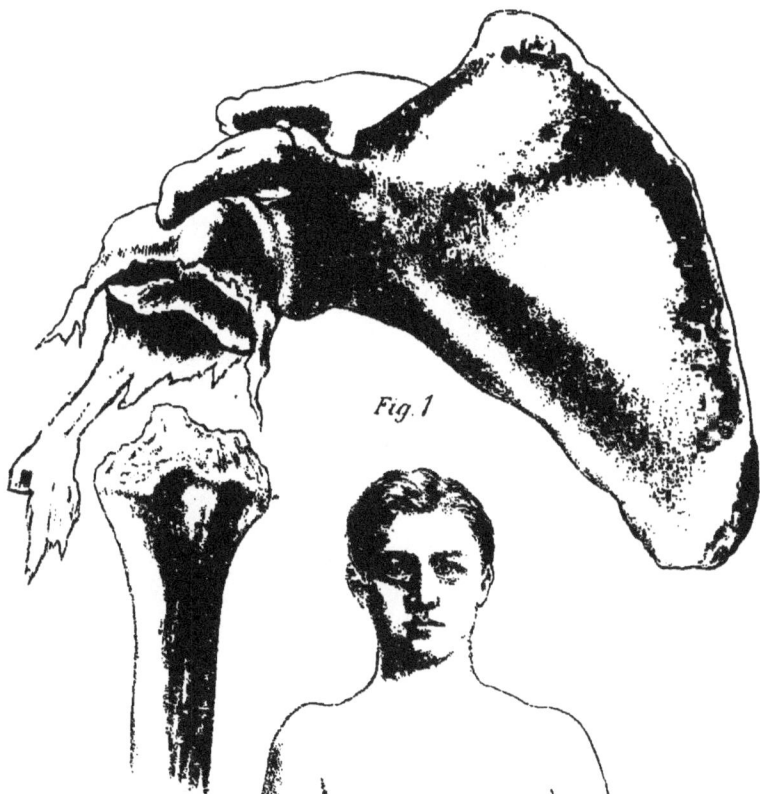

Fig. 1

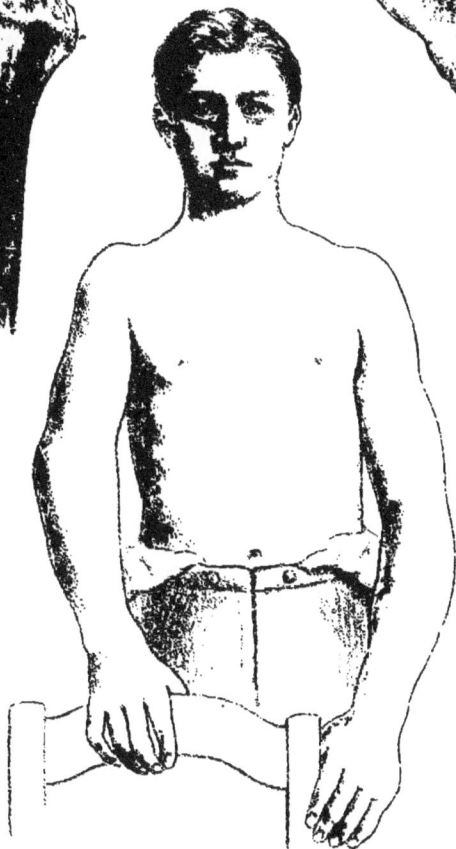

Fig 2

Lith. Anst. v. F Reichhold, München

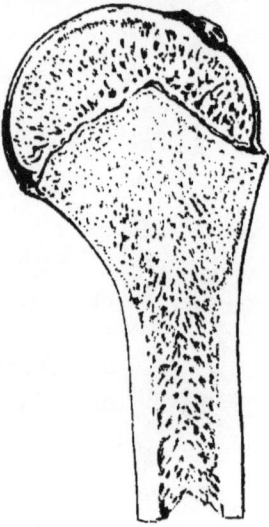

Fig. 1

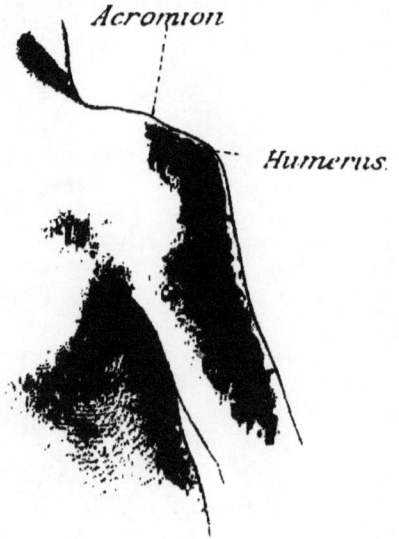

Acromion

Humerus.

Fig 2

Fig 4

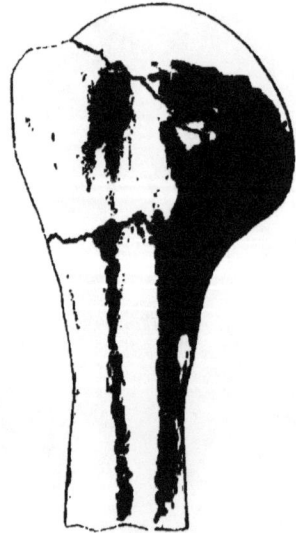

Fig 3

Lith Anst v F Reichhold, Munchen

Explanation of Plate 28.

FIG. 1.—*Course of the epiphyseal line on the section of the normal bone.* The epiphysis at the upper end of the humerus really consists of the epiphysis of the head and the apophyses of the tuberosity. But these coalesce so early into one bone that practically the epiphysis is of importance in this form. The elevation of the epiphyseal line shown in the illustration is the result of the coalescence. The knowledge of this form is of practical importance because not rarely we can discover by direct palpation the diaphyseal end with its pointed extremity and lateral declivity. (Author's collection.)

FIG. 2.—Picture of the shoulder of a boy of 14 (Klinke, 1894), who had suffered the displacement shown by a traumatic separation of the epiphysis at the upper end of the humerus. The illustration is to recall the fact that this displacement is best recognized, not from in front, but, as here shown, from the dorso-lateral aspect. The diagnosis will be easiest when the observer stands behind the patient and takes, as it were, a bird's-eye view, that is from above, inspecting the injured shoulder and at the same time comparing the two sides; in this way we observe the characteristic, sometimes almost angular, projection of the displaced diaphyseal end of the humerus.

FIG. 3.—Lines of fracture drawn in the *anatomical* and the *surgical* neck of the humerus.

FIG. 4.—Old fracture of the upper part of the shaft of the humerus united with marked displacement; the shaft is displaced forward and inward. (Author's collection.)

Explanation of Plate 29.

Fractures in the Middle of the Humerus.

Fig. 1.—Anatomical preparation of the arm to show the *position of the radial nerve with reference to the bone.* We see the yellowish nerve resting directly upon the bone; the brachialis internus is pushed slightly forward, the triceps backward. Below, the elbow can be recognized. The view of the arm is directly from without; and, as is well known, it is at the junction of the lower and middle third of the humerus that the radial nerve can be felt at the outer side, as a rule, during life. It is readily understood that the nerve in this position is very liable to be injured when the humerus is fractured by some trauma. In all injuries affecting the middle and lower third of the humerus the surgeon should always bear this fact in mind; it is very annoying if it become evident at a later period that the patient is unable to raise the dependent hand dorsally (by reason of paralysis of the extensors).

Figs. 2 and 3.—United fractures of the shaft of the humerus with some displacement of the fragments. Fig. 3 is the external view of the same bone shown in section on Plate 6, Fig. 2. (Author's collection.)

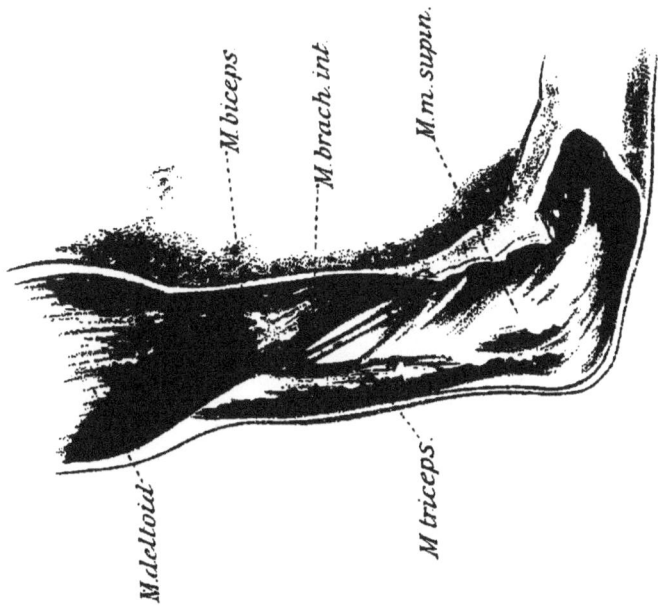

M.deltoid.

M biceps.

M. brach. int.

M.m. supin.

M triceps.

Fig. 1

Fig. 2

Fig 3

Fig. 1ᵃ

Fig 2

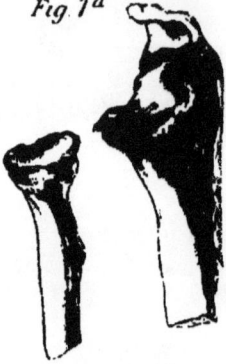

Fig. 1ᵇ

Fig 3

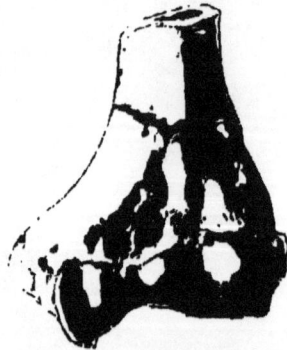

Fig 4

Explanation of Plate 30.

FIG. 1 *a* and *b*.—Bones of a child (right arm) injured by heavy machinery. In Fig. 1 *a* we see the transverse fracture and the fissure in the shaft of the humerus, also the *partial separation* of the inner and middle portion *of the lower epiphysis of the humerus.* The forearm bones of the some patient are represented in Fig. 1 *b;* the radius is normal; the ulna shows a longitudinal fracture causing separation of the olecranon. The arm had to be amputated. (Author's collection.)

FIG. 2.—*Longitudinal fracture of the humerus extending into the elbow-joint.* The injury was due to a charge of small shot at close range. The humerus at its middle was completely comminuted; the lower portion exhibited the longitudinal fracture shown. The patient recovered with the loss of the arm. (Author's collection.)

FIG. 3.—Oblique fracture through the articular end of the humerus with *separation of the eminentia capitata and the external condyle.* Such and similar oblique fractures occur in all possible variations. (Author's collection.)

FIG. 4.—Typical transverse fracture of the humerus above the condyles, with longitudinal fracture extending into the joint, so-called T-fracture. (Author's collection.)

Explanation of Plate 31.

FRACTURES AT THE LOWER END OF THE HUMERUS AND AT THE CAPITULUM OF THE RADIUS.

FIGS. 1 and 2 show the epiphyseal line and the epiphysis at the lower end of the humerus in section and on external view. This traumatic separation of the epiphysis in its pure form is much rarer than at the upper end of the humerus.

FIG. 3.—*Fracture at the lower end of the humerus above the condyles, with typical displacement.* An attentive observer will immediately recognize the similarity of the position and of the posterior outline of the arm to that in posterior luxation of the forearm (Plate 34). The bones are indicated by dotted lines, and the conditions thus made evident in the illustration are recognized by the physician by careful palpation; he finds the olecranon less markedly projecting, the condyles which are easily felt are in normal relation to the forearm bones, and on displacing the lower end of the humerus laterally the point of fracture is readily discovered by the crepitation. The treatment of course must secure the replacement of the fragments, not rarely and most suitably by the aid of extension by weights and pulleys. (Personal observation.)

FIG. 4 *a* and *b.*—*Old united fracture of the capitulum of the radius.* The cartilaginous edge has a rather tumid character; the separated fragment has united in a displaced position, which is particularly marked in the section (Fig. 4 *b*). The specimen was obtained by resection (Friederike Lemke, aged 28, 1889. Three months before admission she had fallen on her outstretched arm).

38

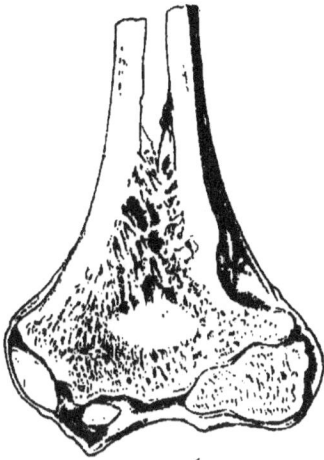

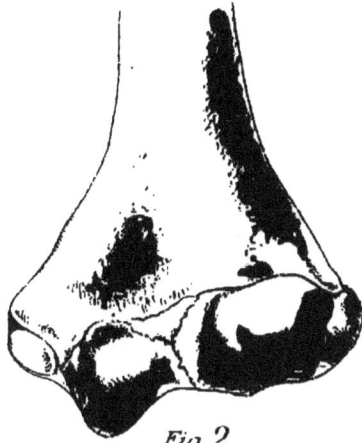

Fig. 1

Fig. 2

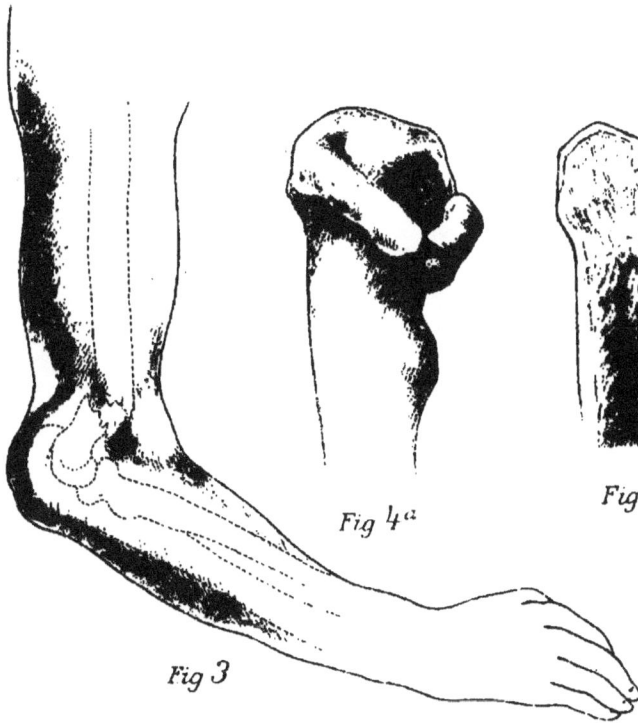

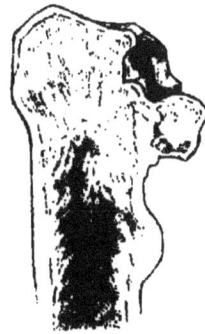

Fig 4ᵃ

Fig. 4ᵇ

Fig 3

Lith. Anst .v. F. Reichhold, München.

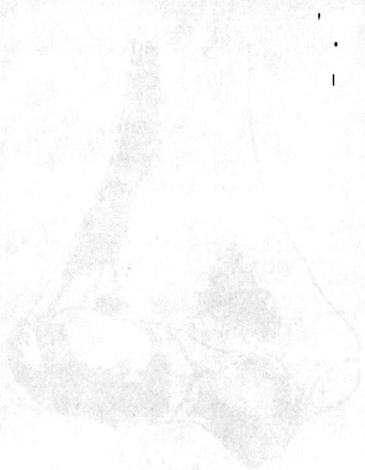

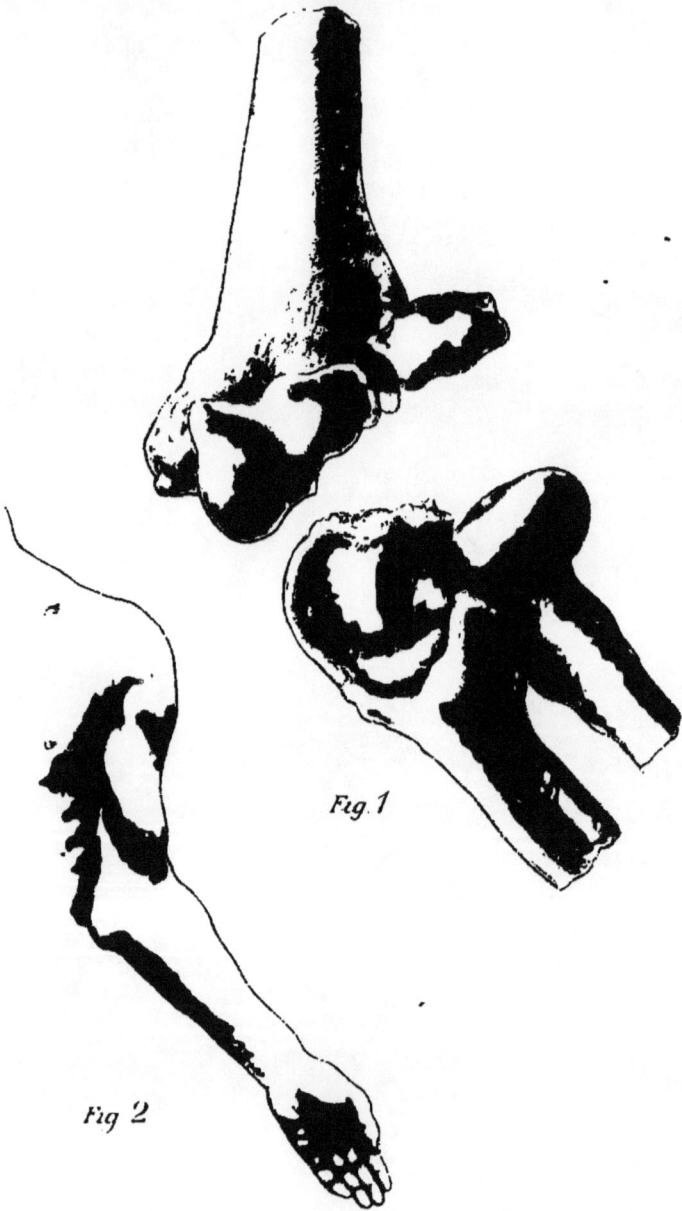

Fig. 1

Fig 2

Lith Anst v F Reichhold, München

Explanation of Plate 32.

DEFORMITY OF THE ARM AFTER ARTICULAR FRACTURE AT THE LOWER END OF THE HUMERUS.

FIGS. 1 and 2.—*Old oblique fracture at the lower end of the humerus, with the formation of a cubitus valgus.* The osseous preparation shown in Fig. 1 is an accidental finding in a cadaver. We observe the alterations following this old articular fracture and presenting the character of a slight grade of arthritis deformans, namely, tumid thickening of the capitulum of the radius, atrophic conditions of the articular ends with their cartilaginous covering, sparse thickening of the bone in their neighborhood. (Author's collection.)

In Fig. 2 the same condition may be recognized in the living patient. This was a man, aged 34 (John Janker, 1884, Surgical Policlinic, Munich, No. 1,140), whose deformity was caused by a fracture two years previously, which had united obliquely. The illustration was drawn from a photograph.

As we meet with genu valgum and genu varum, which occasionally result from trauma and intra-articular fracture, so we may observe also a cubitus varus or valgus after an unfortunate fracture at the lower end of the humerus. In every hinge joint such displacement and termination are possible after a separation or an oblique fracture. The measures by which such results may be avoided are exact reposition of the fragments and keeping them in good position either by splints, the arm being flexed or extended, or by extension by weights and pulleys in straight or flexed position, possibly by the aid of lateral weighting of the fragments (sand bags) or suitable lateral traction. Frequent inspection will be required.

Explanation of Plate 33.

OUTWARD LUXATION OF THE FOREARM, WITH SEPARATION OF THE INTERNAL CONDYLE.

Of the lateral luxations of the forearm the outward is more frequent than the inward variety and is usually associated with separation of the internal condyle. The strong internal lateral ligament is not torn even by a powerful force (in an abduction movement at the elbow as in the production of a valgus position), but the bone is often separated.

FIG. 1 shows the lateral displacement of the bones so that the ulna articulates with the lateral part of the trochlea and on the eminentia capitata, while the capitulum of the radius projects free. The separated internal condyle is still connected with the ulna by the internal lateral ligament. The drawing was made from an artificial preparation in the cadaver.

FIG. 2.—In this illustration the luxation shown in Fig. 1 is easily recognized. The contours of the arm are otherwise little changed; especially is there an absence of the projection of the olecranon characteristic of backward luxation. The prominence of the capitulum of the radius is obvious on inspection alone. Palpation would confirm this finding, especially when movements of pronation and supination are made. (Personal observation.)

40

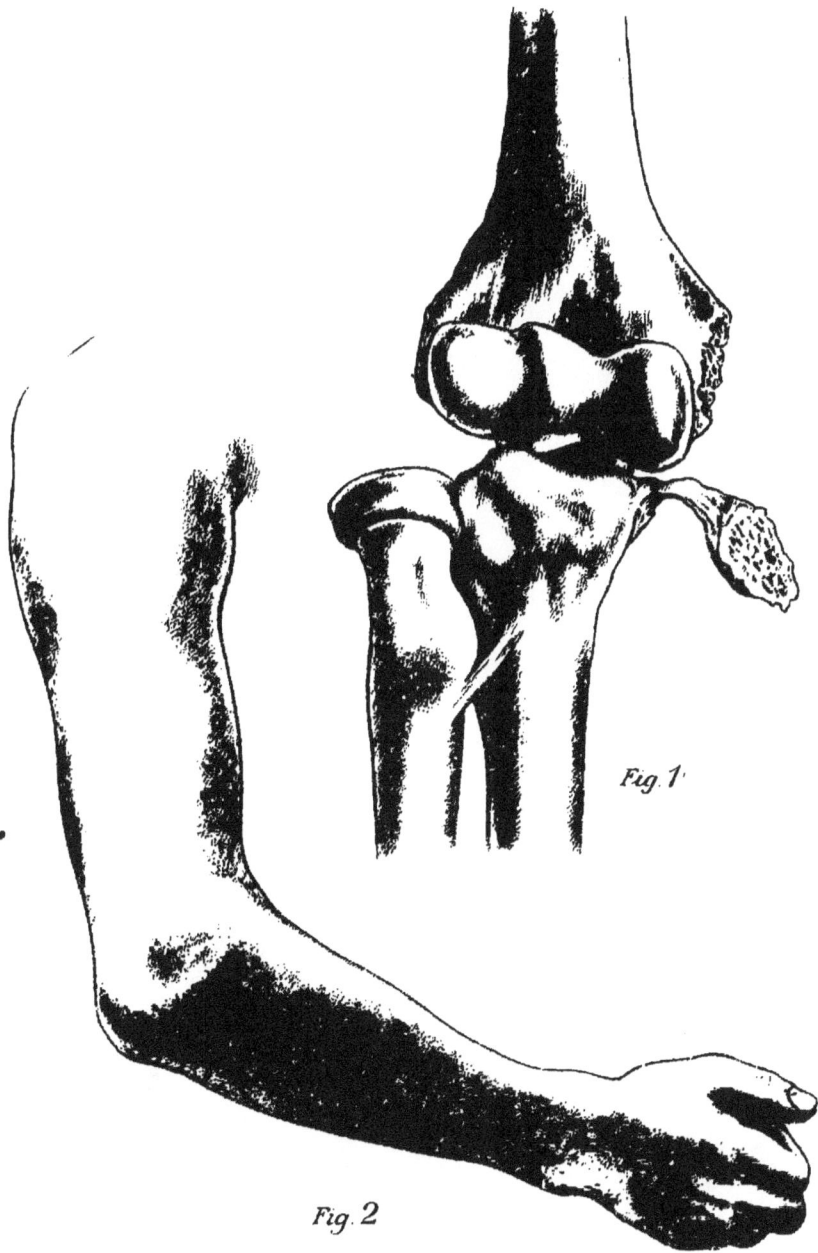

Fig. 1

Fig. 2

Lith. Anst .v. F. Reichhold, München .

Lith Anst v F Reichhold, Munchen

Explanation of Plate 34.

BACKWARD LUXATION OF THE FOREARM.

The drawing was made from nature after preparations on the cadaver. The position shown (about at a right angle) on the one hand was particularly appropriate for representing the details, and on the other hand it was rendered necessary by the size of the small plates. *In this injury, as a rule, the arm occupies a more obtuse angle.*

The minute anatomical details are made very clear by Fig. 1. We see the shaft of the humerus and its lower articular end; beneath it and abnormally displaced backward the capitulum of the radius and the articular cavity (cavitas sigmoidea majora) at the upper end of the ulna. Very interesting here is the delineation of the external lateral and the annular ligament, which is strictly true to nature. At the anterior side the biceps with its tendon, beneath it one margin of the brachialis internus, behind the humerus the triceps with its insertion at the point of the olecranon can be recognized.

FIG. 2 shows this form of luxation in the living patient. We recognize the rounded, turban-like prominence of the capitulum of the radius, which, together with the olecranon, projects abnormally backward. When we form a mental picture of the longitudinal axis of the humerus we become at once aware that it does not coincide below (as under normal conditions) with the articular end of the forearm bones, but is divided into a short posterior and a long anterior portion. Corresponding to this is the characteristic alteration of the posterior contour.

41

Explanation of Plate 35.

REDUCTION OF POSTERIOR LUXATION OF THE FOREARM BY HYPEREXTENSION AND TRACTION.

An attentive observer will have noticed in Fig. 1, Plate 34, the relation of the coronoid process of the ulna to the humerus. In this luxation, owing to the tension of the ligaments and muscles, there is naturally a very close apposition of the dislocated bones to the humerus; the coronoid process is often in the posterior supratrochlear fossa, where it is almost incarcerated. For this reason an attempt to force the reduction by simple longitudinal traction is faulty. The incarceration must first be freed by hyperextension in the elbow-joints, as shown in Fig. 2. To this end the patient is as a rule anæsthetized and a position is effected as in Fig. 1. During this act the hand applied to the elbow can at the same time follow the result of the hyperextension and the effect of the traction which succeeds. The bones of the forearm slip over the margin of the articular surface of the humerus into their normal position, and the reduction is completed by the flexion at the elbow which follows. The illustration was drawn from a photograph of this method of reduction of the luxated arm.

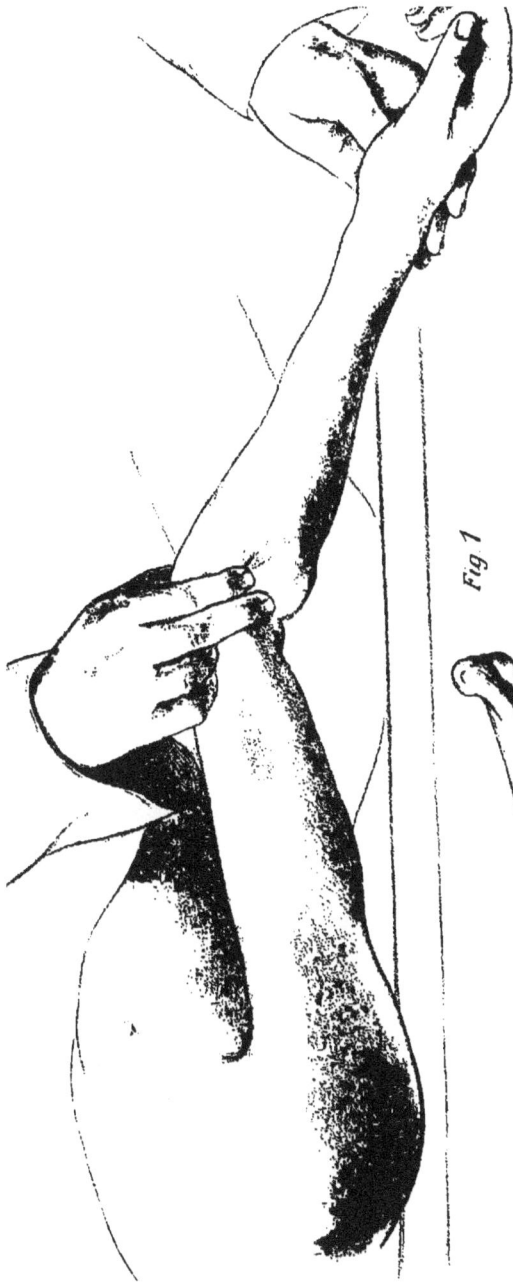

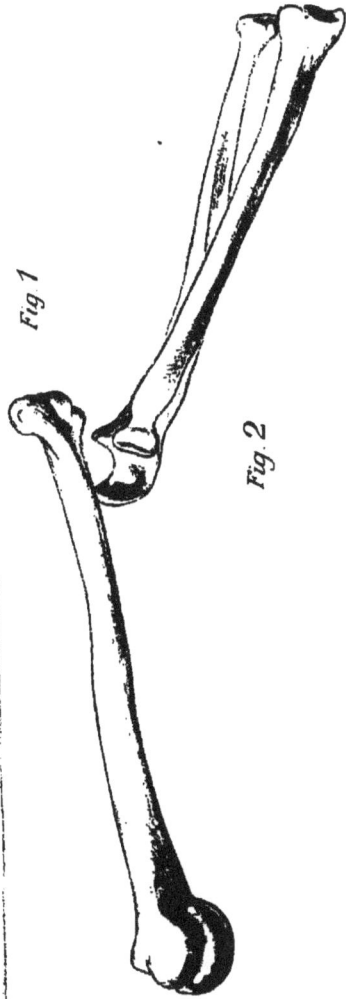

Fig 1

Fig. 2

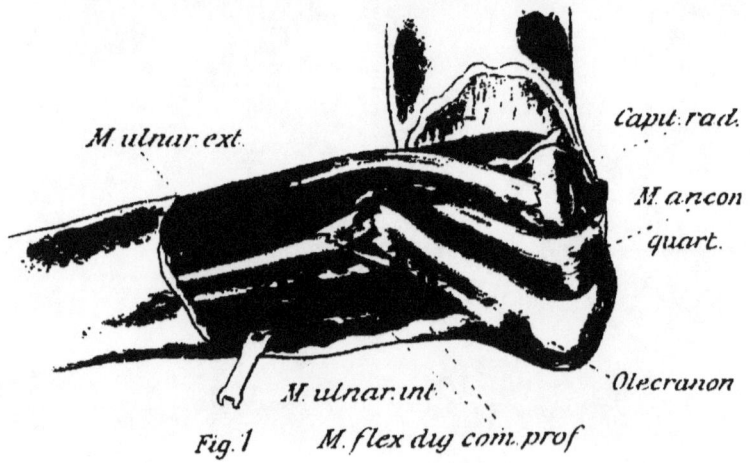

M. ulnar. ext.

Capit. rad.

M. ancon
quart.

M. ulnar. int.

Olecranon

Fig. 1

M. flex. dig. com. prof.

Fig 2

Lith Anst v F Reichhold, München

Explanation of Plate 36.

FIG. 1 shows the more minute anatomical details of this typical injury, as they appear in an artificial preparation. The ulna with its fragment displaced at a distinct angle strikes the eye at once; the capitulum of the radius is also readily recognized. Between the head of the radius and the olecranon the anconæus quartus muscle shows very clearly; below the ulna we see the flexor digitorum communis profundus and the ulnaris internus; above the ulna the ulnaris externus.

FIG. 2.—The same injury, which was artificially produced in the cadaver, drawn by the aid of a photograph from a case observed during life. The position of the arm, the angular flexion in the upper portion of the ulna, and the prominence of the capitulum of the radius are characteristic. Close behind and below the head of the radius the projection of the external condyle can be made out. This injury is not rarely misunderstood, and when neglected can no longer be thoroughly and completely cured. It requires osteotomy at the seat of the ulnar fracture and operative reduction or resection of the head of the radius. It is hoped that this illustration will contribute toward a better understanding of this typical injury.

43

Explanation of Plate 37.

FIG. 1.—*Fracture of the olecranon.* The drawing was made from an artificial preparation in the
cadaver. We see the ulna with the separated olecranon and the diastasis between the two fragments
which during life is effected by the traction of the
triceps. Of course the position adds to the effect,
since the diastasis of the fragments is considerable
when the elbow is flexed, and can usually be overcome
when the arm is completely extended. The olecranon is at the same time slightly detached or twisted.
It is very clearly evident that a fracture of the olecranon cannot exist without a wide opening of the
joint; the cartilaginous surface of the lower end of
the humerus is free; at this point of course we find
the effusion of blood which results from the injury.

Looking at this illustration with a view to the
treatment, it is quite evident that the first requirement in fracture of the olecranon is extension of the
arm at the elbow-joint, aspiration of the blood if the
effusion is large, and an approximation of the upper
fragment by traction effected by means of a strip of
adhesive plaster applied in the form of a sling.

FIG. 2.—Old bone preparation of a *fracture of the
olecranon* healed by ligamentous instead of osseous
union. (Author's collection.)

FIG. 3.—Illustration of a *separation of the coronoid process.*

44

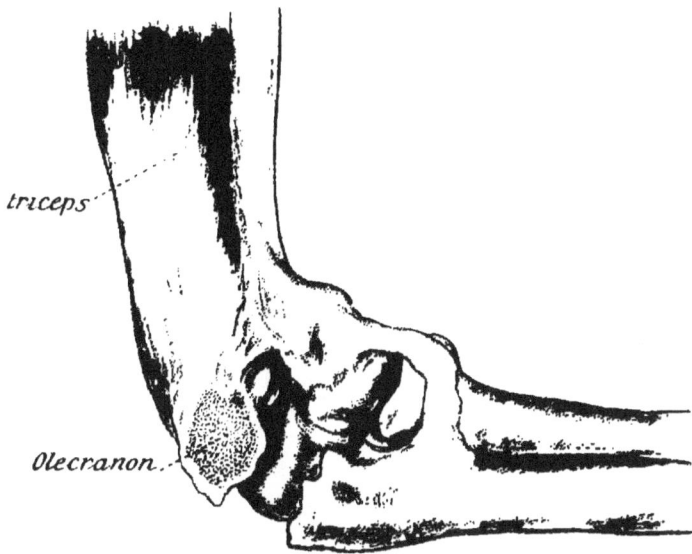

triceps

Olecranon

Fig. 1

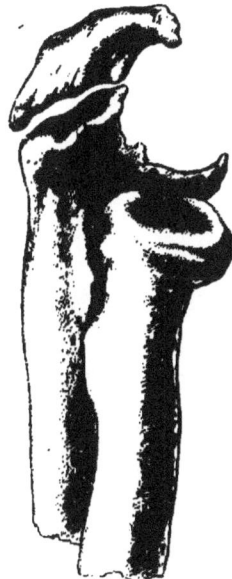

Fig. 3

Fig. 2

Lith. Anst .v. F. Reichhold, München

Fig 1

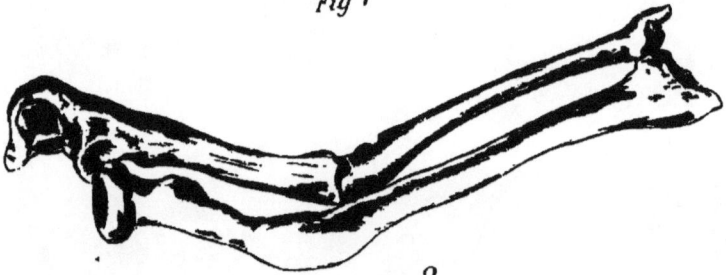

Fig 2

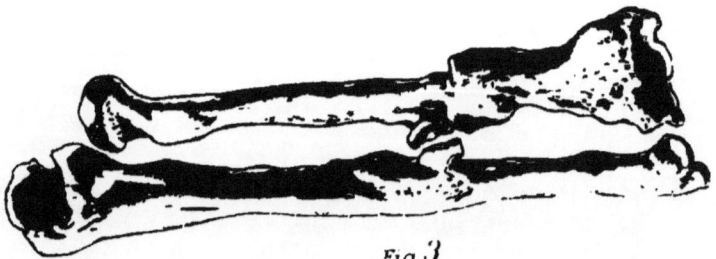

Fig 3

Explanation of Plate 38.

FIG. 1.—The illustration shows the *unfavorable position of the fragments* in fractures of the middle of the forearm which presents itself not rarely in recent fractures, and now and then in old fractures where osseous union failed to occur and a pseudarthrosis resulted. The drawing was made from nature, the patient being a boy who recently came under treatment. In this case the fragments could be replaced under anæsthesia, and very good union was obtained by extension of the arm in the elbow and careful fixation by means of a long dorsal splint.

FIG. 2 shows a similar angular position, but the injury is older; *the radius has undergone firm bony union*, while the *ulna* is still movable, in a condition of *pseudarthrosis;* both bones are bent in an equal degree. The result shown in the specimen was probably due to insufficient fixation of the fracture, possibly to a dressing which was too short and failed to include the two neighboring joints, a blunder often committed by quacks. (Author's collection.)

FIG. 3 shows a most important condition in a relatively harmless form: *the two bones are connected at the point of fracture*, fortunately not by a mass of bone but in the form of a *nearthrosis;* from each bone springs a conical projection at whose point is a kind of joint surface which articulates with that of the other bone. It is clear that a faulty position of the two bones (pronation) or the pressure of a firm circular dressing may cause the coalescence of the callus of the two bones, when their fracture is in a corresponding position. (Author's collection.)

Explanation of Plate 39.

FIG. 1.—*Isolated fracture of the radius above
its middle and the effect of the biceps on the
position of the upper fragment.* In this illustration, which is drawn exactly from nature (artificial
preparation), we see the forearm with the hand and a
portion of the arm. The forearm is in pronation.
The upper fragment of the radius, however, under the
influence of the biceps, is in supination; for this muscle, as is well known, produces supination and flexion
of the supinated forearm. We recognize the supination of this upper fragment by the position of the
tuberosity of the radius (the point of insertion of the
biceps) and especially by careful inspection of the line
of fracture: the lower fragment shows on the fractured
surface a small inferior defect caused by the formation of a dentation at the upper fragment; the dentation and the defect do not face, but the former, by
outward rotation of the upper fragment, *i.e.*, by its
supination, has been displaced nearly 180°. In view
of these facts it is evident that even in isolated fracture of the radius *the arm must be dressed in the
supine position.*

FIG. 2.—*Epiphyseal lines at the lower end of
the radius and ulna,* after a dry preparation. True
separation of the epiphyses of these bones, especially
of the radius, is not rare in children. As a consequence of such injuries I have observed serious disturbances of growth. (Author's collection.)

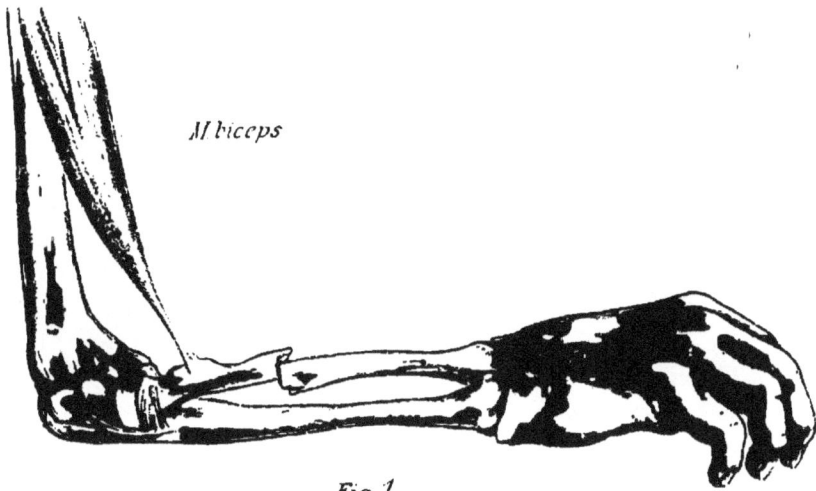

M. biceps

Fig. 1

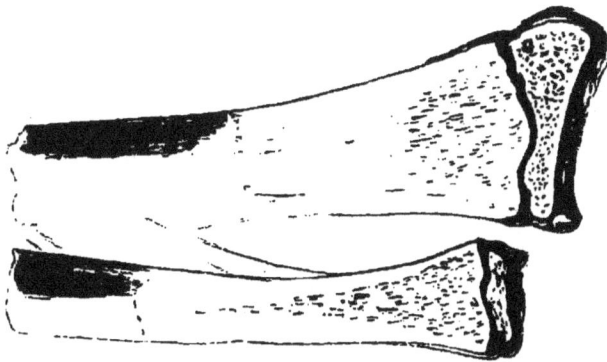

Fig 2

Lith. Anst .v. F Reichhold, München

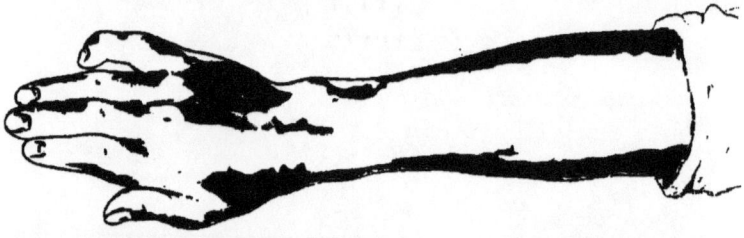

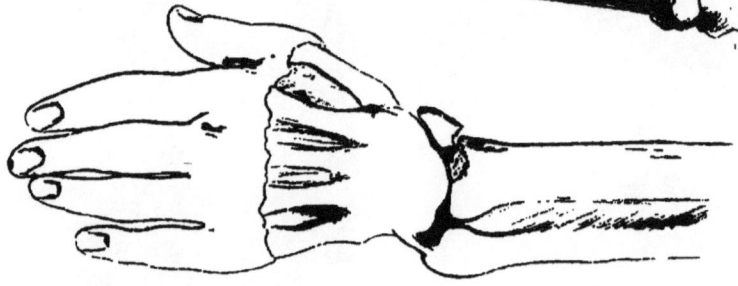

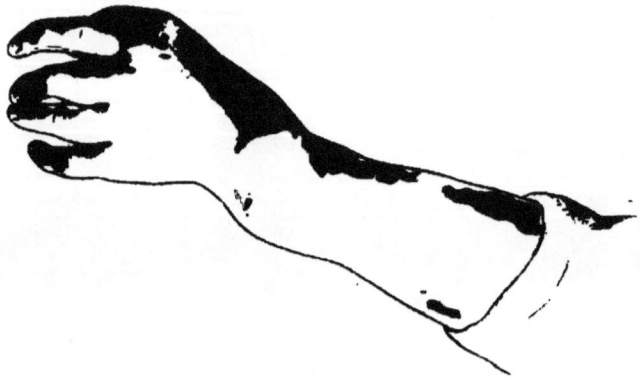

Fig 3

Explanation of Plate 40.

TYPICAL FRACTURE OF THE LOWER EPIPHYSIS OF THE RADIUS.

Fracture at the lower end of the radius is one of the most important of all fractures. This injury is still much sinned against, both by incorrect treatment of the rightly diagnosticated fracture and by its misinterpretation and neglected replacement. The diagnosis of this typical fracture can generally be made by inspection, first from the dorsal side, as on Plate 40, and then laterally, best from the radial side, as on Plate 41.

During this inspection the healthy and the injured side should be carefully compared. FIG. 1 *a* and *b* therefore shows the healthy right arm and the deformed left arm of the same patient. The physician had best be seated opposite the patient, who is likewise sitting. He then will perceive, as in Fig. 1 *b*, a marked radial displacement of the whole hand in the region of the wrist-joint, with a consequent prominence of the styloid process of the ulna, and a widening of the forearm in the region of the styloid process.

FIG. 2 shows an artificial fracture of the radius, only a portion of the articular surface being separated; this variety of fracture is frequently observed, as are other oblique fractures in various directions up to a pure transverse fracture in about the old epiphyseal line. The prominence of the ulna and the radial displacement of the hand are here also visible; the upward displacement of the fragment, that is, a rotation around the lower articular surface of the ulna as a centre, could not be made apparent.

FIG. 3 represents a transverse fracture of both forearm bones after a photograph (Mina Houdelet, aged 60, 1890), in which the deviation of the fragments may also be recognized.

47

Explanation of Plate 41.

Typical Fracture of the Lower Epiphysis of the Radius.

This plate represents a typical fracture of the epiphysis of the radius in side view. In Fig. 1 we recognize the displacement of the fragments which French surgeons call "en dos de fourchette" (silver-fork, back-door, or Colles' fracture). The forearm at its lower end shows, instead of a slight dorsal protrusion, a marked prominence on the flexor side, corresponding to the angular break at the seat of the fracture. The longitudinal section in Fig. 2 makes this very clear. There is a transverse fracture of the entire lower epiphyseal portion of the radius, with separation of small marginal pieces, in the classical displacement of the fragments. A bayonet-like position results, since the longitudinal axis of the radius and the hand is interrupted and step-like, owing to the oblique direction of the fragment. The anatomical details are apparent from the lettering on the plate.

The preparation was artificially produced by forced hyperextension of the hand, with strong pressure upon the latter in the direction of the forearm. It is clear that replacement will often be suitably begun by forced flexion of the hand. It is not surprising that this fracture was formerly regarded as a luxation at the wrist-joint. Now we know that this luxation is among the rarest injuries.

48

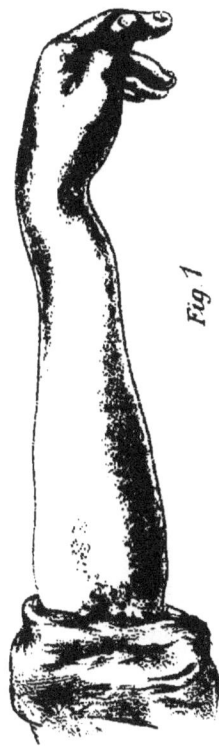

Fig 1

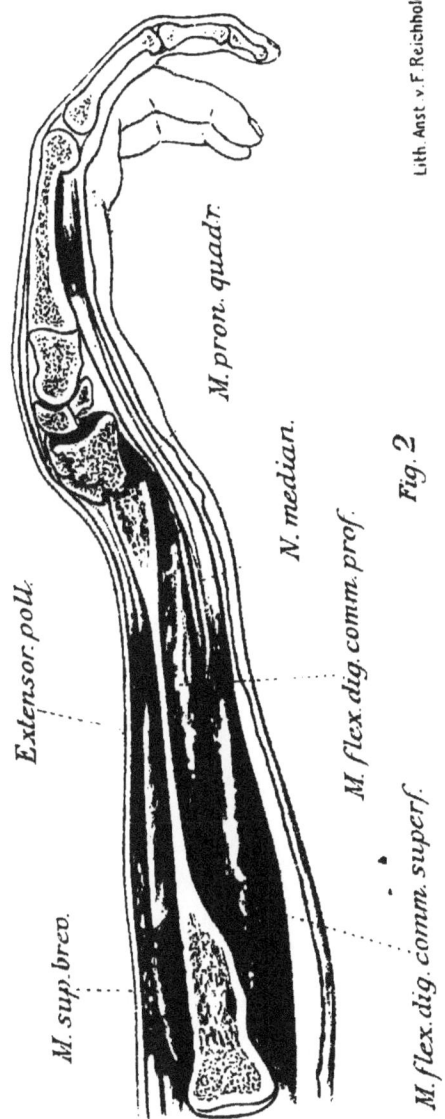

Extensor. poll.

M. pron. quadr.

N. median.

Fig. 2

M. flex. dig. comm. prof.

M. sup. brev.

M. flex. dig. comm. superf.

Lith. Anst. v. F. Reichhold, München

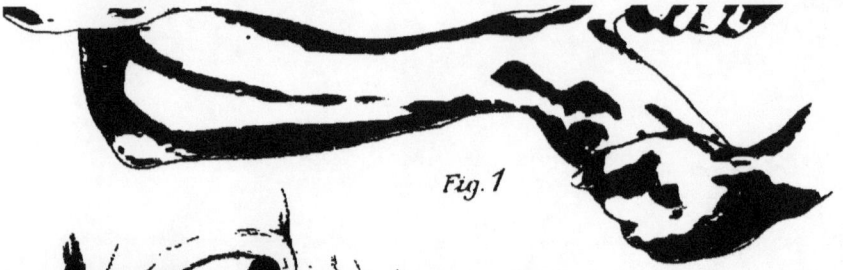

Fig. 1

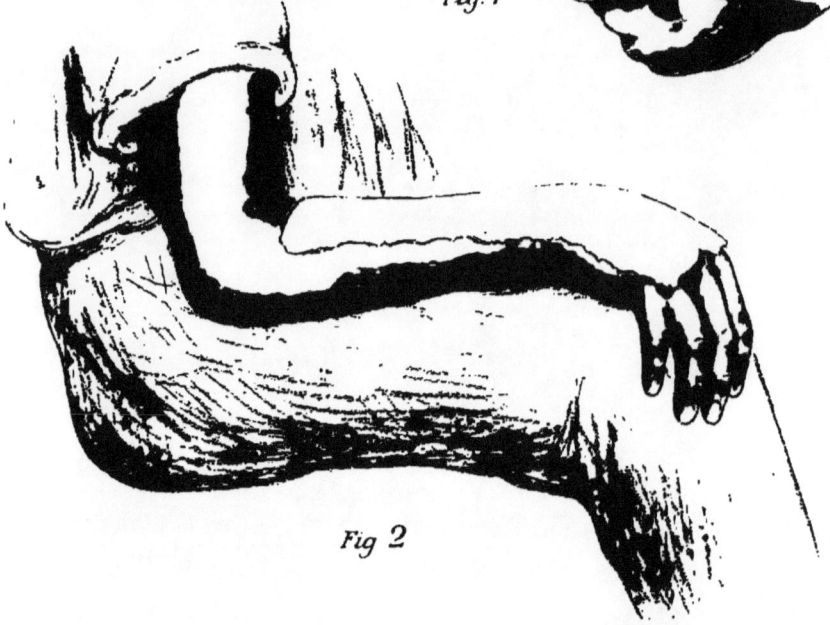

Fig 2

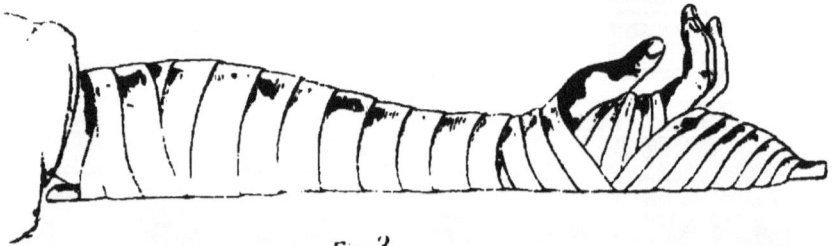

Fig 3

Lith Anst v F Reichhold, Munchen

Explanation of Plate 42.

FIG. 1, drawn from a photograph of the patient, shows how the surgeon is assisted during the *replacement of the fragments in typical fracture of the epiphysis of the radius.* One assistant pulls upon the thumb and fingers of the injured hand, in the manner shown, the other assistant exerts counter-pressure on the arm. The surgeon then can place the fragments in the desired position by direct pressure upon the seat of the fracture.

FIG. 2.—*Application of a Beely's plaster-of-Paris splint* after replacement is effected. The forearm may suitably rest upon the patient's thigh, the hand being flexed in the volar and ulnar direction. In this position the application of a plaster-of-Paris splint with hemp or jute fibres is easily made. *But no splint should project beyond the metacarpus; the fingers must remain free.* (After a photograph.)

FIG. 3.—*Illustration of the dressing devised by Professor Roser.* The forearm and hand in complete supination rest upon the wooden splint so padded that the hand is in volar flexion, the fingers again being free.

Recently Professor Petersen has urged that the arm after replacement of the fragment should be simply placed in a mitella, the hand being pendulous, and that splints should be dispensed with. In many cases this can certainly be done and it may be interesting to know that the same thing was proposed some time ago by Hutchinson ("Illustrations of Clinical Surgery," II., 110). In the majority of cases, however, this will not be feasible in medical practice, if only for the reason that it requires very careful and frequent examination.

49

Explanation of Plate 43.

TYPICAL LUXATION OF THE THUMB.

Luxation of the thumb is practically much more important than most text-books would lead one to believe. The present plate shows the anatomical relations of this luxation. It appears that the first phalanx is dislocated upon the dorsal surface of the first metacarpal; the head of this bone is readily recognized, surrounded by the muscles inserted at the first phalanx. We see that the head of the first metacarpal projects markedly; the capsule is lacerated on the flexor side and is dislocated dorsally with the first phalanx. With reference to the ulnar and radial sides of the head of the metacarpal, we recognize at once that the adductor pollicis and the tendon of the flexor pollicis longus lie on the ulnar side, the flexor pollicis brevis and the abductor pollicis brevis on the radial side. The capitulum lies as in a slit between these muscles, and especially close to the tendon which completely surrounds the neck of the metacarpal and is hidden behind it, but is again visible on the volar side of the first phalanx. The illustration was drawn from nature after a preparation in which the thumb was artificially luxated.

50

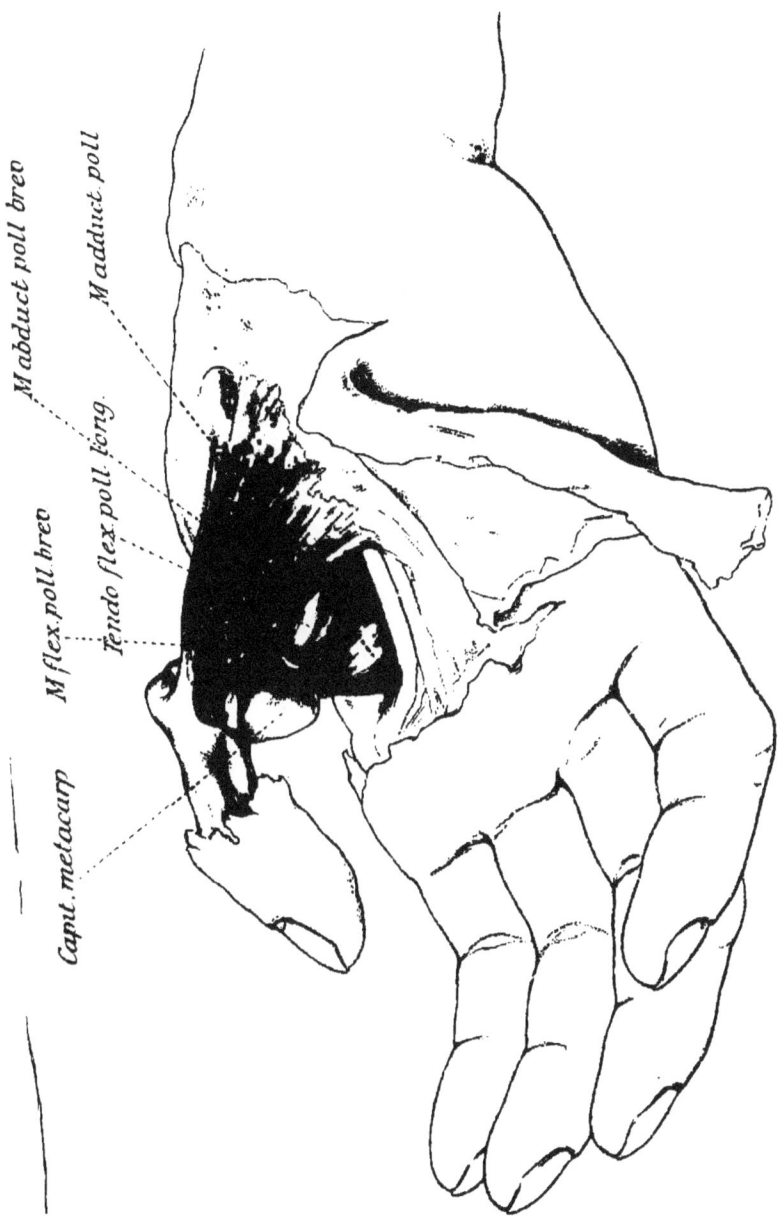

M abduct poll brev

M adduct poll

Tendo flex poll long.

M flex. poll brev

Capit. metacarp

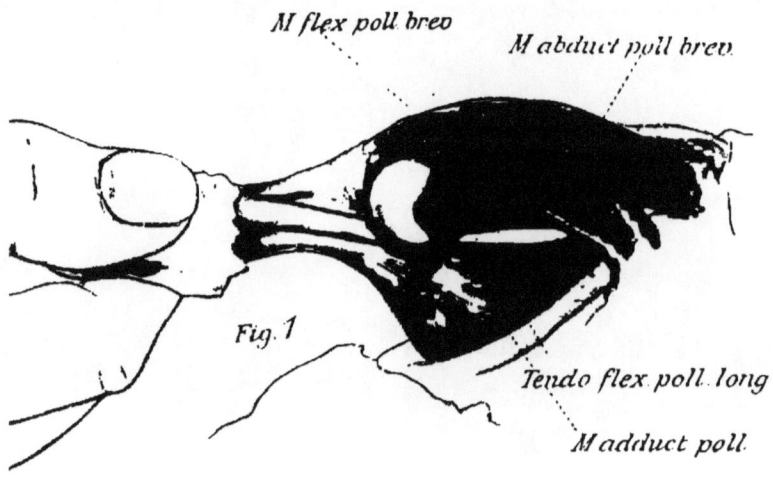

M flex. poll. brev.

M abduct. poll. brev.

Fig. 1

Tendo flex. poll. long.

M. adduct. poll.

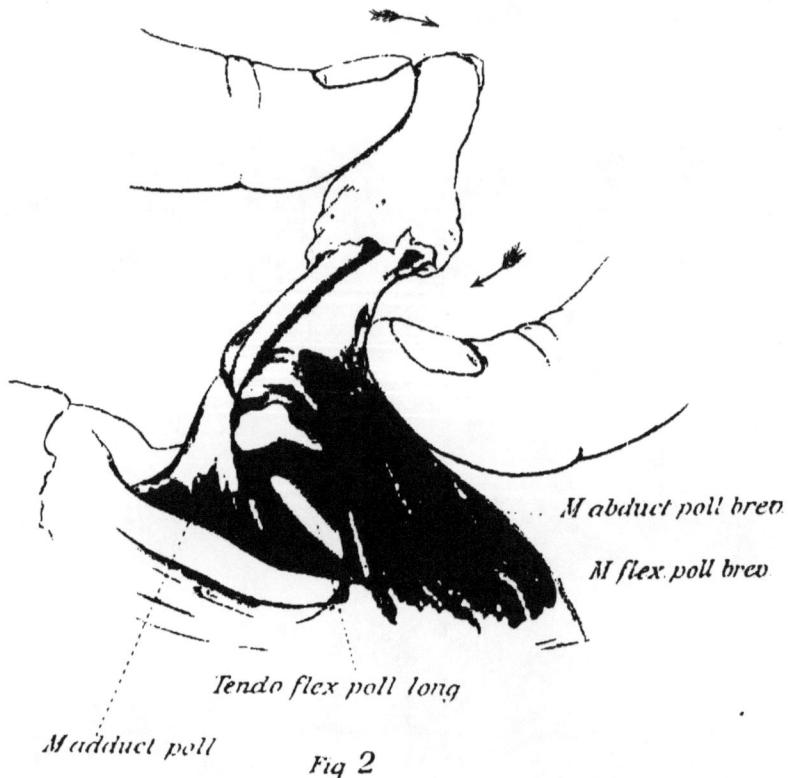

M. abduct. poll. brev.

M. flex. poll. brev.

Tendo flex. poll. long.

M. adduct. poll.

Fig. 2

Lith. Anst. v. F. Reichhold, München

Explanation of Plate 44.

The reduction of the typical luxation of the thumb was in former times often performed quite *incorrectly,* and false statements and drawings are still found in many books. As in analogous luxations of all hinge joints no force should be employed; forceps such as were formerly used with a view to exert powerful traction must be altogether rejected. *Any traction renders the reduction more difficult,* for while it continues the muscles and the tendon of the flexor pollicis longus hug the neck of the metacarpal head and form a real obstruction to reduction. This is partly shown in Fig. 1; the long tendon surrounds the bone and at the same time is slightly turned on edge.

The *correct mode of reduction* is represented in Fig. 2. First hyperextension must be effected, as indicated in Fig. 2, by finger I. pressing in the direction of the arrow. The thumb in this position then is *pushed forward* along the base of the first phalanx, as it were crowded beyond the head of the metacarpal, as shown by Finger II. pushing in the direction of the arrow. This manipulation succeeds in reducing simple cases if it is performed with the necessary skill. It fails only when special conditions result from an interposition which will become apparent and be overcome by making a longitudinal incision in the direction of the projecting metacarpal head and separating the tissues until the obstruction is reached.

51

IV. Fractures and Luxations of the Upper Extremity.

INJURIES of the upper extremity may have a direct or an indirect causation. While a direct force produces certain lesions whose presence can frequently be inferred from a knowledge of the cause, indirect injuries of different forms may be due to one and the same cause. Thus a fall on the hand may produce a typical fracture at the lower end of the radius, an injury in the elbow-joint, at the upper end of the humerus, or in the shoulder-joint, and in children frequently enough a fracture of the clavicle.

1. CLAVICLE.

Fractures of the clavicle may affect any part of the bone, but are most frequent about the middle. The symptoms of this typical fracture of the clavicle, in the large majority of cases of an indirect origin, are as a rule characteristic. The displacement of the fragments is due both to muscular traction and to the weight of the arm. The sternal fragment is in-fluenced by the sterno-cleido-mastoid and is usually displaced slightly upward. Owing to the traction of the powerful muscles passing from the thorax to the arm, the external fragment with the whole arm is ap-proximated to the thorax; for under normal conditions the clavicle acts as it were as a cross-beam which

49

keeps the region of the shoulder away from the thorax. As a consequence of these relations, in typical clavicular fracture, the arm sinks down; it is lower than on the healthy side. Secondly, the arm as a whole is approximated to the thorax and consequently the axilla is obliterated. Thirdly, the arm is displaced forward and inward, a kind of inward rotation, obviously the result of the predominant traction of the thoracic muscles.

The diagnosis of clavicular fracture therefore is very simple, especially because the displaced fragments can be felt directly on this superficial bone, and the pain and the functional disturbance point to the seat of the injury.

The treatment of these typical clavicular fractures requires in the first place a very accurate replacement, and then a dressing which will constantly counteract the causes of the displacement. As is well known, it was formerly considered a rare thing and an almost impossible task to cure such a fracture without displacement of the fragments. Our present auxiliaries enable us to effect recovery almost invariably in good position, even in severe cases of this nature.

During the replacement and the application of the dressing (Fig. 2) it is advisable to have an assistant stand behind the seated patient and draw both shoulders of the latter vigorously backward. For a dressing the strips of adhesive plaster recommended by Sayre are suitable. Three such strips are required, two of which serve for correcting the displacement mentioned above. The first strip corrects the inward rotation of the arm or shoulder region; it passes at the upper end of the arm from within outward over the

shoulder to the back. The second strip lifts the depressed arm by passing from the elbow region to the
healthy shoulder. The third strip acts merely as a
mitella parva; it raises the hand and passes to the
injured shoulder, while it at the same time exerts a

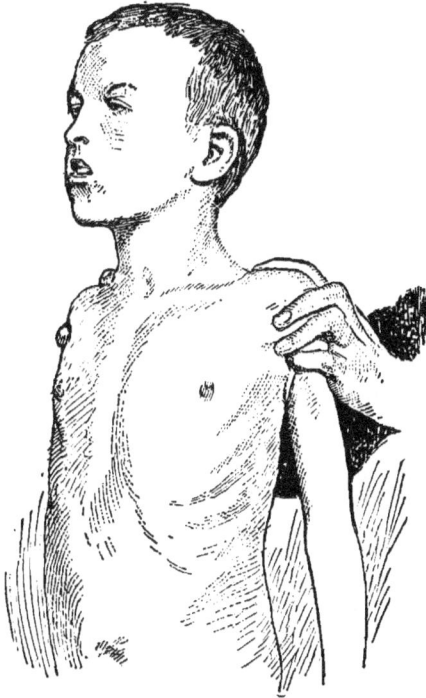

FIG. 2.

gentle pressure from in front and above upon the fragments. One indication which is not quite fulfilled by
this dressing is the restoration of the axillary cavity.
This indication is met by placing into the axilla a
well-fitting cushion of some soft material (cotton or
wood wool wrapped in mull) and retaining it there.

The effect of this dressing is strengthened by a few turns of a roller bandage. In applying this, occasionally a small pad may be so fixed over the fracture as to exert slight pressure from above upon the sternal fragment. In summer it is desirable to dust the parts to be covered by the dressing, especially the axilla, with some toilet powder.

In order to intensify the effect of these strips of adhesive plaster it might be useful to insert pieces of rubber bandage in the strips and place them at such tension as to exert continuous elastic pressure which would counteract a tendency to a recurrence of the displacement. The tension could be regulated by an appropriate application of a rubber tube. With proper supervision and if the surgeon possesses the necessary technical skill he will succeed in securing satisfactory results.

Incidental injuries may involve the brachial plexus and more rarely the large vessels. A portion of the plexus may also be injured secondarily by pressure of the callus from which it cannot escape owing to its position on the first rib.

Fracture in the median and lateral third of the clavicle as a rule is not associated with displacement, excepting fracture at the extreme acromial end, in which the lateral fragment often is almost upright. Cases of separation of a splinter and infraction are to be treated on the above principles.

LUXATIONS OF THE CLAVICLE.

a. Sternal luxation, *i.e.*, dislocation of the sternal end of the clavicle, occurs in different varieties, namely:

Forward (presternal).

Upward (suprasternal.) These occur only indirectly through leverage when the first rib serves as a fulcrum, or through an extraneous force according to

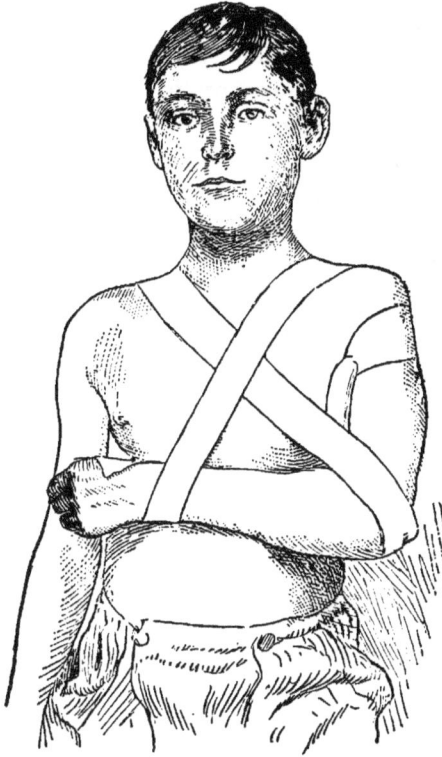

FIG. 3.

the position of the clavicle, backward or downward. As regards the former variety, secondary displacement may likewise be of importance.

Backward (retrosternal), very rare, through direct force.

The diagnosis is always easy because all parts are accessible to palpation. In backward luxation respiration and deglutition may be interfered with through pressure on the trachea and œsophagus. In differentiating fractures near the articular end use is made of palpation of the normal rounded bony prominences and mensuration of the length of the clavicle.

Treatment.—Replacement is generally easy; retention, that is, maintenance of the correct position, difficult. The requirements are exact dressings, with direct pressure on the replaced articular extremity, sometimes those having an elastic effect (see Treatment of Clavicular Fractures), occasionally fixation by means of percutaneous suture.

b. Acromial luxation, namely :

Upward (supra-acromial).

Downward (infra-acromial), the latter very rare.

The former often results from direct force acting upon the acromion when the clavicle is fixed; it is therefore really a downward luxation of the scapula. This luxation is complete when the dislocation is extensive owing to rupture of the coraco-clavicular ligament.

The diagnosis is easy, since exact palpation may be made, but this dislocation is sometimes mistaken for a luxation of the humerus. (Compare the description of Plate 25.)

Treatment.—Here, too, replacement is easy and retention often very difficult. By turns of a roller bandage the arm is elevated and the clavicle at the same time pressed down. Sometimes an elastic bandage or the percutaneous suture of the ligaments will be required (Baum).

2. SCAPULA.

Different forms of fracture of the scapula occur; those of the body and the spine of the scapula are direct and are often associated with several lines of fracture and fissures, though the fragments are but slightly displaced. Crepitation and abnormal mobility can often be felt, especially if the arm be in a suitable position. Treatment: fixation of the arm.

Fractures at the neck of the scapula are rare, but are most liable to occur at the surgical neck, that is to say, the coracoid process remains attached to the articular portion and the line of fracture extends downward from the notch (see Plate 24). This fracture of the neck of the scapula is important in differential diagnosis, since it may be mistaken for subcoracoid luxation of the humerus. The symptoms of this fracture are descent of the arm, which may even be slightly abducted, and marked prominence of the acromion; the deformity disappears with crepitation when the arm is elevated, but returns immediately when the arm is released; sometimes the edge of the fractured surface can be felt from the axilla. Union generally results from the employment of a dressing which puts the arm and scapula at rest and fixes the arm in its correct position by the use of an axillary pad, in a way similar to Sayre's adhesive-plaster dressing in clavicular fracture. The arm must be permanently elevated and kept slightly outward and backward.

Separation at the margin of the socket, especially of the inferior portion, is rare and can be recognized

as an intra-articular injury only in certain positions
of the arm at the shoulder-joint. There is some descent of the head of the humerus when the arm is
kept in a lateral horizontal position, and at times
crepitation when the head of the humerus is moved
from before backward. Isolated fractures of the
coracoid process, by direct force, are exceedingly rare;
those of the acromion are more frequent and are
diagnosticated by direct palpation and the demonstration of abnormal mobility and crepitation; sometimes
the fissure may be felt when the arm is vigorously
drawn across the body. Union results when the arm
is slightly elevated and placed at rest.

3. SHOULDER-JOINT.

Luxations at the shoulder-joint are among the most
important and frequent injuries. Their diagnosis is
usually not difficult, and still some cases pass un-
recognized. On the normal shoulder we feel the
acromion extending from the spine of the scapula, its
connection with the clavicle, below it the coracoid
process, then the head of the humerus under the deltoid muscle usually so distinctly that on its rotation
we can even palpate the tuberosity and the intertuber-
cular sulcus, and from the axilla the head of the humerus and the margin of the glenoid fossa. As is
well known, this very movable joint is kept in contact, not by the capsule and ligaments, but by the
muscles and atmospheric pressure. In paralysis of
the deltoid muscle the head of the humerus always
sinks slightly, and there are cases of essential paralysis in children in which the descent of the head of the

humerus is at once perceptible through the thin over-lying soft parts.

a. Forward luxation of the humerus (preglenoid, subcoracoid, or subclavicular, according to the degree of dislocation of the head under the coracoid process or the clavicle) is the most frequent luxation at the shoulder-joint. The artificial production of this dislocation in the cadaver resting on its back is usually easy when the arm is highly elevated laterally or abducted and gradually but vigorously pressed backward. In this way the capsule is greatly stretched antero-internally by the advancing head, it gives way (this is its thinnest portion), the head passes through the lacerated capsule forward under the coracoid process, and the luxation is complete; as soon as the arm is placed in a more normal position all the objective signs excepting the effusion of blood are present.

During life subcoracoid luxation sometimes arises directly from a postero-lateral blow against the humerus, more frequently indirectly by a fall on the side while the arm is raised and abducted, or else by a fall upon the extended hand or the elbow, especially when the arm is directed backward. The luxation has also been observed as a result of violent movements of the arm (hurling, flinging), that is, through muscular action.

In the indirect occurrence of this luxation, by excessive abduction the arm at last comes into lateral contact with the scapula; the region of the tuberosity and the surgical neck of the humerus press against the upper margin of the glenoid fossa if the force continues to act; the latter point forms a fulcrum and

the short lever, *i.e.*, the head of the humerus, is lifted
out of its normal position and connections. The luxation when thus effected is as a rule more or less
downward, infraglenoid; but by a secondary displacement of the humerus (muscular traction) the
subcoracoid variety results.

The symptoms of the typical subcoracoid luxation
are quite characteristic. They are all based on the
fact that the head of the humerus is absent from its
normal position and present in an abnormal location.
The examination always begins with inspection,
which alone often suffices for the diagnosis, so that
palpation is needed merely to place it beyond doubt.
It is best that the patient be seated free on a chair,
his trunk being bare, so that the physician sitting
opposite can easily inspect and compare both sides.

The rounded outline of the shoulder has disappeared and the acromion forms an angular projection. The normal round outline of the shoulder is
formed by the head of the humerus and the deltoid
muscle; if the latter is much atrophied the acromion
is prominent; if the head of the humerus is absent
from its normal position the acromion presents an
angular projection even when the deltoid is well developed and in spite of the effusion of blood which
exists. That this prominence is of the acromion is
easily determined by following the spine of the scapula, which terminates in the acromion, from the
back.

In the region below the coracoid process an abnormal prominence is visible and palpable; it is felt particularly if the humerus is slightly turned forward
and backward, when it becomes evident that the

prominence is part of the arm, and its rounded form proves it to be the head of the humerus.

The arm is in a position of elastic abduction, that is to say, it can be adducted into contact with the thorax by moderate force, but on being released it immediately springs back into the abducted position. This symptom is caused by the tension of some ligaments (the coraco-humeral) and the muscles inserted at the tuberosity.

The longitudinal axis of the arm passes to the coracoid process or under the clavicle, instead of under the acromion, its normal direction. This fact is ascertained especially by comparison with the healthy side.

The external contour of the arm forms an angle open outward, while on the healthy arm it is nearly straight. This change is the result of the abducted position of the arm to which the lower half of this outline corresponds, and of the tension of the fibres of the deltoid between the acromion and the arm, which fibres form the upper portion of this angular contour.

The humerus seems to be elongated and the distance from the acromion to a point at the elbow (for instance, the external condyle of the humerus) is actually very often lengthened; it certainly is never shortened. This is manifest also on inspecting the patient from behind. The explanation of this elongation is furnished at once by producing this luxation in the skeleton: the head of the humerus is really somewhat below its normal position in the socket. It is self-evident that during mensuration and inspection the two arms should be exactly symmetrical in position.

At the same time the head of the humerus can be felt more or less distinctly from the axilla in its false position; passive movements are painful and limited, and active movements are still more restricted.

Among incidental injuries to be observed may be mentioned separation of splinters about the major tuberosity, rarely lesions of the vessels, but more often nerve lesions. In this luxation the nerves are always tensely stretched, sometimes they are bruised during the occurrence of the luxation by the head of the humerus or compressed between the latter and the thorax. The axillary nerve is particularly liable to be injured, and therefore it is desirable to test the deltoid muscle supplied by it, immediately after replacement lest an error be made in the prognosis.

The diagnosis of this luxation, therefore, is not difficult as a rule; should any doubt remain it will be removed by examination under anæsthesia. In the differential diagnosis the following conditions are to be considered :

Contusion of the shoulder and distorsion of the shoulder-joint present no dislocation.

Supra-acromial luxation of the clavicle. In this accident the angular projection is caused by the acromial end of the clavicle and not by the acromion. The arm is not abducted.

Fracture of the neck of the scapula. The acromion projects, the head of the humerus is lowered and slightly displaced forward and inward, but the dislocation disappears when the arm is simply pushed up, during which act crepitation is usually felt.

Paralysis of the deltoid muscle is followed by descent of the arm, but this displacement disappears at

once when the arm is pushed up. Moreover, the arm
is not abducted.

Fracture of the acromion with marked displacement of the fragments. In this injury the anatomical relation between the acromion and the head of the
humerus remains unchanged.

Fracture of the neck of the humerus. The round-
ness of the shoulder is preserved even when the fragment of the shaft is displaced inward; the arm is not
elastically abducted and is never lengthened; on the
contrary it is nearly always shortened.

Treatment.—An early reduction should unquestionably be attempted. By the exercise of some skill
the reduction may be effected without resorting to
anæsthesia; often the head is easily replaced while the
surgeon pretends to make merely an exact examination. In other cases the attempts fail, when anæsthesia should be at once induced.

Of the many methods of reduction which have been
devised and performed in the course of time only the
following need be recommended here:

1. Extension by an assistant of the slightly abducted arm of the recumbent patient, counter-extension being effected by means of a broad cloth wound
about the thorax. The surgeon at the same time
performs manipulations, especially direct pressure
upon the head toward the socket. (Cooper's well-
known method: Traction upon the arm in the longitudinal direction of the body, with simultaneous
pressure of the foot [without shoe] in the axilla, thus
exerting direct pressure upon the head.)

2. Kocher's method of rotation. This consists of
several steps or positions which must be exactly fol-

lowed (compare Plate 22). The order of the steps is,
adduction of the arm until it touches the trunk; outward rotation until the flexed forearm is about in the
frontal plane (to be done with great care lest fracture
result); forward elevation of the arm; and finally
inward rotation.

It is particularly by this method that reduction
succeeds not rarely without anæsthesia and in the
most gentle manner. The adduction causes tension
of the upper wall of the capsule and fixation of the
head at the margin of the capsule, so that during rotation it turns about the latter and not on its axis.
The elevation relaxes the coraco-humeral ligament.

The success of the reduction is indicated by a more
or less distinct snap of the head and by the restoration of the normal shape and the regained mobility.

In the after-treatment the arm is best so fixed by
cloths, bandages, or strips of adhesive plaster that
the hand of the injured side rests on the healthy
shoulder. After a week passive movements are begun, which are followed later on by active movements. The whole duration of the treatment until
the patient can return to work is about four to five
weeks.

If reduction fails, other attempts must be made
under profound anæsthesia, the laceration of the capsule having been enlarged by extensive movements.
If these attempts fail, even with the aid of other
physicians, operative reduction must undoubtedly
be performed in order to force restitution as soon as
possible. By using the incision for resection, from
the coracoid process downward, success will be easily
obtained.

If the reduction is not effected, the result will be as a rule the most undesirable condition of an old luxation. But rarely a nearthrosis with some mobility is formed; generally the region of the shoulder remains painful and the movements are reduced to a minimum. Even in such old cases improvement may be secured by arthrotomy and reduction or by resection. In rare cases a state of habitual luxation develops.

Modifications and Complications of Præglenoid Luxation.

When the head of the humerus leaves the socket directly forward, it lies between the scapula and the subscapular muscle, often so close to the socket that the articular surface of the head still touches the margin of the glenoid fossa. In such cases, which mainly result from direct force, the bones abrade each other at their points of contact within a few weeks. In old cases of this kind the abrasion is pronounced, being in the form of a deep groove at the head of the humerus, and of an erosion of the anterior half of the glenoid fossa; at the same time, however, we find the well-known periosteal proliferations by which a kind of new socket is formed for the head in its abnormal position. (Compare Plate 23.) The reduction of such cases is usually very difficult, and often impossible without arthrotomy.

Supracoracoid luxation is extremely rare and always associated with fracture of the coracoid process.

a. Luxation combined with fracture of the neck of the humerus is a serious injury. When reduction by traction aided by direct manipulation fails, arthrotomy must be performed and reduction forced. Formerly it was advised to secure a false joint at the seat of the fracture, leaving the head in the luxated position.

b. Downward (infraglenoid or axillary) luxation of the humerus. In this dislocation the head is at the lower margin of the glenoid fossa and can be at once felt from the axilla. When the arm is elevated horizontally the appearance is quite characteristic, a marked bayonet-like depression of the line of the shoulder being visible. The acromion is prominent, the articular cavity is empty, and the function is disturbed. Sometimes the arm is raised (luxatio erecta) or fixed horizontally. Reduction succeeds by traction on the arm and direct pressure against the head from the axilla (the thumbs if need be pressed against the acromion).

c. Backward (retroglenoid, subacromial, infraspinata) luxation of the humerus is very rare and usually due to direct force. The head is easily seen and felt in its abnormal position; the coracoid process projects markedly. Reduction is effected by traction upon the arm with adduction and direct pressure.

4. ARM.

A. FRACTURES AT THE UPPER END.

a. Fracture of the anatomical neck (Plate 28, Fig. 3) is very rare, especially in a pure form. Should only that portion of the head which is covered with cartilage break off, a strictly intracapsular fracture, the vitality of the fragment would be doubtful; it would act like separate osseo-cartilaginous fragments, for instance, in the knee-joint. Hence as a rule this fracture is not purely intracapsular, but the fragment is generally attached to and nourished by portions of the capsule.

The cause is usually a direct force. The head may be impacted between the tuberosities, or the upper diaphyseal end in the cancellous structure of the head; the displacement may be very slight or very great, the fragment of the shaft being shifted forward, inward, and upward; complete inversion of the fragment of the head has also been observed, the cartilaginous surface being turned toward the fractured surface of the shaft of the humerus.

The symptoms are those of an intra-articular injury. The diagnosis is possible only by careful examination under anæsthesia, with deep palpation of the bony points and the demonstration of abnormal mobility and crepitation of the upper end of the humerus.

Treatment.—Rest in bed with extension of the arm downward and outward by weights, a pillow being inserted in the axilla or additional traction employed, acting laterally at the upper end of the humerus. Passive movements should be begun early.

b. Fracture at the surgical neck (Plate 28, Fig. 3) is a frequent injury. The line of fracture lies below the tuberosity or penetrates into it. The upper fragment therefore may still be under the influence of the muscles inserted at the tuberosity. The fracture is usually due to a fall upon the shoulder, in persons of somewhat advanced age; sometimes it results from a fall upon the hand or the elbow. The fragments may be fixed by impaction; compare the specimen of

a fracture by compression, Plate 3, Fig. 2. The
fragments may also unite with marked displacement,
the upper end of the shaft of the humerus being frequently shifted forward, inward, and upward.

Symptoms.—On palpating the lateral contour of
the shoulder we perceive under the acromion the
spherical head of the humerus in its normal position
and the injured arm adjoins the thorax (is not in
elastic abduction); in many cases the arm is also
shortened (which will distinguish it from the subcoracoid luxation). Usually abnormal mobility and
crepitation (especially on rotating the arm) are also
present; sometimes the before-mentioned displacement of the end of the shaft forward, inward, and
upward can be demonstrated. In the latter case
there is some similarity to subcoracoid luxation, but
the shortening and the other symptoms enumerated
above serve for the differentiation. When the fragments are impacted the diagnosis may become more
difficult, but exclusion of a luxation will always be
possible. With reference to a fracture associated
with luxation, *vide supra* under Luxation.

Treatment.—Reposition is to be carefully effected
when a marked displacement is present. In the further treatment splint dressings for fixing the whole
arm with the shoulder region as far as the neck, together with the use of an axillary cushion, are sufficient unless there is a tendency to a displacement of
the fragments. In that event it is better not to rely
on ambulatory treatment of this serious injury which
ultimately may become incurable through functional
disturbance; instead, we may employ rest in bed and
permanent extension by weight and pulley in the lon-

gitudinal direction of the arm, together with an
axillary cushion or, better, a second extension apparatus acting upon the upper end of the shaft. During this treatment the region of the shoulder which
is bare is open to inspection, massage can be employed, and (the weights being temporarily removed)
we may begin in the first few days with careful passive movements. Otherwise all that has been said
above in the general part about the treatment of
fractures involving the joints applies to this injury.

c. Fracture of the tuberosity, especially the major
one, during the occurrence of subcoracoid luxation,
that is, always combined with the latter (as a fracture by traction). The lesser tuberosity is even less
apt to break. The symptoms of an isolated fracture
of these bony processes are not pronounced; we find
the signs of a contusion and sometimes are able to
feel the mobility of the fractured osseous prominence. This fracture may also occur during the reduction of old luxations of the shoulder.

Transtubercular fracture means a transverse fracture of the humerus at the level of the tuberosity;
compare the remarks upon fracture of the surgical
neck.

d. Traumatic separation of the epiphysis at the
upper end of the humerus (Plates 27 and 28). This
injury is of great practical importance on account of
its relative frequency. Of course it is possible only
before the ossification of the so-called epiphyseal cartilage (better, intermediary cartilage), that is to say,
in young persons, and is due to a fall upon the
shoulder or the arm.

In order to understand the lesion it is necessary to

know the anatomical details of the epiphyseal line;
compare the explanation of Plate 28, Fig. 1.

The symptoms are often quite characteristic; they
point to a separation of bone, as in fracture at the surgical neck. When the displacement is moderate it
may sometimes be possible to demonstrate under
anæsthesia abnormal mobility and crepitation, but
the latter is softer than the ordinary—cartilage crepitation. Not rarely, however, the displacement is
considerable, the diaphyseal end being shifted forward and inward, at which point it sometimes causes
a circumscribed, almost angular prominence which is
most distinct on inspection from the side or from
above (standing behind the patient). In rare cases
the displacement is such that the fragment of the
shaft is almost luxated inward and upward. In that
case reposition even under anæsthesia may be very
difficult or impossible. If it succeeds the further
steps are as in fracture at the surgical neck. If it
fails reposition must be forced by cutting down upon
the parts and separating the interposed soft structures.
I have the records of several similar cases in which
the fixation of the replaced fragments was effected
with excellent results by the insertion of a long steel
needle.

Exact reposition is necessary in order to save these
young people from deformity and functional disturbance persisting through life. Moreover, imperfect
reposition after such lesions of the epiphyseal cartilage is regularly followed by marked arrest of growth:
the humerus does not reach its full length and remains
shorter than the opposite one (Plate 27, Fig. 2).

B. Fractures of the Diaphysis of the Humerus (Plate 29).

These arise directly or indirectly and present the ordinary symptoms of a fracture in a manner readily demonstrated, namely, abnormal mobility, crepitation, various degrees of displacement, etc. In fractures below the attachment of the deltoid muscle the latter may lift the upper fragment outward (dislocatio ad axin). In fractures in the region where the middle and lower thirds of the humerus join, the radial nerve is liable to be involved, primarily by lesion during the occurrence of the injury or secondarily by the pressure of the callus in which it is often imbedded as in a deep groove. This should be borne in mind from the start (paralysis of the extensors of the hand, impossibility of flexing the hand dorsally), lest a gross blunder be committed in making the prognosis. Lesions of the vessels are of rarer occurrence.

Union is normal under correct treatment. But the occurrence of a pseudarthrosis is relatively more frequent after fractures of the humerus than after those of the other bones of the upper extremity. This is due to a more difficult immobilization and to the displacement which is sometimes considerable, and in addition may be complicated by the interposition of soft parts between the fractured ends.

Treatment.—In applying a circular dressing including the region of the shoulder and the elbow joint the axilla must be protected from injurious pressure. Plaster-of-Paris or wire splints or padded tin splints (if the latter are used, a long piece should cover the

whole outer surface and a shorter piece the inner
surface of the arm) are suitable for the dressing.
Wire splints can be applied without difficulty in such
a way as to cause permanent traction in the longitudinal direction of the arm.   The splint is bent accordingly and firmly fastened to the forearm flexed at
a right angle; the upper end is so curved as not to
hug the shoulder closely.   If then a well-padded loop
of bandage is placed in the axilla and fastened moderately tightly to the projecting end of the splint, permanent traction is exerted which can be easily
regulated by renewed tying of the axillary bandage.
Occasionally this dressing may be employed also in
fractures at the upper end and especially at the lower
end of the humerus.

C. Fractures at the Lower End of the Humerus (Plates 30, 31, 32).

These fractures are frequent and of great practical
importance, for in their gravity, though not always
in their anatomical character, they should be looked
upon as articular fractures.   The forms of these fractures are illustrated in the plates: Supracondylar fracture, T-fracture, fracture in the epiphyseal line,
oblique fractures through the articular extremity,
and isolated fracture of the internal and external
condyle.

In all these injuries very careful palpation is required.   A person who is familiar with the relative
position at the normal elbow of the two condyles and
the olecranon will appreciate the pathological condition in the patient, especially if the healthy and the
injured side are compared.

FIG. 4.

a. Supracondylar and T-Fracture.

The former is generally due to a fall upon the elbow or the hand and is of frequent occurrence in children. The T-fracture, *i.e.*, longitudinal fracture of the lower fragment, is caused by the on-crowding of the olecranon from behind or the shaft of the humerus from above. The displacement which occurs in a typical form by the traction of the triceps and gives rise to confusion with luxation of the forearm may be seen on Plate 31. During the examination an important manipulation is to grasp the lower end of the humerus transversely and firmly at its projecting and easily found condyles and to test whether it is abnormally movable. A fracture at the lower end of the humerus can also be recognized by displacing the forearm backward against the fixed arm; thus abnormal mobility may be ascertained together with crepitation, and reposition may be effected. When the shaft of the humerus has penetrated between the two portions of the lower fragment (in a T-fracture), the entire lower articular end is widened.

Treatment.—Thorough reposition, if necessary under anæsthesia, followed by fixation with splints (padded tin splints at the outer and inner side); the elbow being extended or flexed, whichever position facilitates the retention of the fragments. In adults it may be necessary to apply an adhesive-plaster dressing for permanent extension by weights, with the aid of lateral traction slings or direct weighting with sandbags. In children splints will do, but the fact can hardly be sufficiently emphasized that careful reposition and frequent inspection are of the utmost importance; in the case of such fractures in children I resort to anæsthesia during the first and sometimes during subsequent dressings.

Among incidental injuries lesions of the ulnar nerve are rarer than those of the radial; that such complications require careful treatment is self-evident.

b. *Fractures of the condyles* may be isolated or occur as complications, especially with luxations. The diagnosis is readily made by the displacement and mobility of the fragments. The treatment is simple: enforced rest by dressings, early movements.

c. *Oblique fractures of the articular extremity and separation of the epiphyses.* These are articular fractures of a marked form, not rarely associated with serious displacement of the forearm at the elbow-joint. Minute palpation of the several bony prominences and a test of their mobility lead to the formation of a probable diagnosis, which is best made positive under anæsthesia if there be great swelling and acute suffering. Thus we may frequently feel portions of the fractured surfaces and of the articular extremity. With the requisite knowledge of the normal structures and by comparison with the healthy side it will always be possible to gain a correct conception of the nature of the injury.

The prognosis of these fractures unfortunately is in general less favorable than that of supracondylar fractures; for a displacement of the fragments is only too liable to remain behind and to give rise, at this point, to a limitation of the normal excursion of mobility by irregular bony prominences (osseous check). In children and young persons, it is true, the obstruction may be somewhat reduced and mobility improved in course of time by appropriate exercises and the employment of suitable apparatus (for instance, Krukenberg's pendulum apparatus for the elbow joint, which I allow patients to use at their homes), but complete restoration never occurs in these cases. Lateral deviation of the fragments may also lead to displacement; in this way varus and valgus positions develop at the elbow, cubitus varus and valgus. Two pronounced examples of valgus position will be found on Plate 32.

Treatment.—What has been said about supracondylar and T-fractures applies pretty nearly to these fractures. They require exact reposition by manipulation of the forearm and direct pressure under anæsthesia. Then splint dressing in the most appropriate position, sometimes almost or fully extended, sometimes with flexion at the elbow. The flexible padded tin splints are especially suitable because with every change of the dressing—which should be made every three or four days in the first two weeks, and later on every other day—they can be at once bent to correspond with the necessary change in position.

5. ELBOW-JOINT.

For the examination of luxations at the elbow-joint an accurate knowledge of the contours of the normal joint is indispensable. We palpate the condyles, the olecranon, their mutual relations in various positions of the joint, under the external condyle the capitulum of the radius, which is particularly distinct in pronation and supination of the forearm; in luxated posi-

tions the articular ends can often be felt very distinctly, *e.g.*, the capitulum with its central depression,
the eminentia capitata, the trochlea, the upper end of
the ulna. It is part of an exact examination that we
not only believe we recognize one or another bony
process, but we demonstrate the location of all bony
points in their mutual relation, and know their position even when a portion of them cannot be directly
palpated. The skeleton of an arm should be at hand
when these injuries are studied.

We distinguish the luxation of both forearm bones
and the luxation of one of the bones.

*a. Posterior Luxation of the Forearm (Plates
34, 35).*

No dislocation can be more easily produced in the
cadaver than this, which, by the way, is also frequent
during life. Hyperextension causes a laceration of
the joint capsule on its anterior aspect, at the same
time the olecranon is crowded into the posterior
supratrochlear fossa; when the bones are sufficiently
forced apart and a backward push is given to the
forearm while the elbow is flexed, the luxation is
complete. The arm then is bent at an obtuse angle
in the elbow (on Plate 34 the drawing appears with a
right angle owing to lack of room). Further flexion
is opposed by the pressure of the coronoid process
against the articular extremity of the humerus.

In the living patient this mechanism is unquestionably the most frequent; but it is said that the luxation may also result from hyperflexion and by forced
lateral movement.

The symptoms can be readily understood; the projection of the olecranon is immediately visible. The lower end of the humerus may be indistinctly felt by palpation under the soft parts at the bend of the elbow; only when these parts are extensively lacerated (brachialis internus muscle, nerves, and vessels) can it be felt under the skin, though it may also be visible in a fissure of the skin in compound luxations. The longitudinal axis of the humerus does not strike the forearm at its end as under normal conditions, but in such a way that a small portion of it protrudes backward. The olecranon and the capitulum of the radius are accessible to direct palpation, and can be clearly followed when slight movements of the forearm are made. The condyles are at an abnormal distance from the olecranon; the lower end of the humerus admits of no abnormal movements as in supracondylar fracture. The humerus is not shortened. Forward traction of the forearm does not cause the dislocation to disappear.

The diagnosis may be more difficult when complicating injuries are present, such as fracture of the coronoid process; and associated supracondylar fracture of the humerus has also been observed, as well as fracture of the olecranon. In fracture of the trochlea the forearm with this fragment may be dislocated backward, with luxation of the capitulum of the radius.

The prognosis may be rendered unfavorable by complications; otherwise complete passive and active mobility must be secured after reduction.

Treatment.—The mode of reduction is illustrated on Plate 35. As in all hinge joints reduction cannot be affected by simple traction, no matter how powerful. The setting must proceed without any force, as a rule under anæsthesia. The arm is first hyperextended in order to free the coronoid process from its incarceration in the posterior supratrochlear fossa. Then moderate traction at the forearm brings it forward, while the other hand grasps the affected elbow region laterally and controls the position. The flexion now following meets with no opposition; the dislocation has disappeared and the normal contact of the joint has been restored.

The after-treatment is typical: two weeks' fixation with repeated change of the dressings and massage; then mobilization.

b. Lateral Luxation of the Forearm (Plate 33).

Lateral luxations at the elbow joint likewise are not rare; the outward variety is somewhat more frequent than the inward, and is generally associated with fracture of the condyle. This fracture results from traction of the lateral ligament and affects the condyle which is at a greater distance from the forearm; that is to say, outward luxation is complicated with separation of the internal condyle and *vice versa*.

The forearm, however, is generally still in contact, though it be an abnormal one, with the humerus, *e.g.*, the ulna with the eminentia capitata in outward luxation; the capitulum of the radius projects free supero-externally. As a rule the forearm is at the same time displaced backward so as to present a combination of lateral and backward luxation. While backward luxation may occur with intact lateral ligaments (although the internal ligament is usually

lacerated), lateral luxation is in most cases associated
with extensive laceration of ligaments or, as stated
before, with fracture of a condyle.

The form described has also been called incomplete
luxation, as opposed to the complete dislocation of the
bones, in which no portion of one articular surface
remains in contact with a portion of the other.

Lateral luxation always depends upon a•lateral
flexion at the elbow. The capsule is always extensively lacerated, also from the side.

The symptoms of a complete lateral luxation, say
outward, are unmistakable and need no explanation.

In incomplete outward luxation (Plate 33) the
prominence of the capitulum of the radius is clear to
the eye and to the palpating finger. ·On the inner
side the trochlea can be partly felt; the internal condyle may be recognized as detached or appears as a
marked prominence. On making gentle movements
(examination under anæsthesia) the whole condition
becomes clear.

In incomplete inward luxation the external condyle
projects markedly or is broken off; on the inside the
ulna is prominent and its articular surface is palpable; the capitulum of the radius is upon the trochlea;
the eminentia capitata can be partly palpated.

The prognosis depends upon the complications.

Treatment.—The reduction is effected in the most
gentle manner under anæsthesia, by means of hyperextension and lateral pressure with the other hand,
followed by traction and flexion. When interposition
is present, extensive lateral movements (hyperextension with abduction, etc.) are sometimes useful.
Should reduction fail the obstruction must be re-

moved by incision (arthrotomy, best lateral); in this way excellent results may be obtained.

c. *Anterior luxation of the forearm* (the forearm being dislocated forward). This is a very rare injury, whose occurrence was formerly denied unless associated with fracture of the olecranon. This luxation may result from a push or fall upon the olecranon while the forearm is in extreme flexion at the elbow.

Symptoms.—The prominence of the olecranon is absent in its normal position, and the form of the lower end of the humerus can be palpated on its dorsal surface. When the outer surface of the olecranon is still in contact with the trochlea (the arm being nearly straight), the luxation is incomplete; when complete the summit of the olecranon is situated before the articular surface of the lower end of the humerus (the arm being bent at an acute angle). Reduction by direct pressure under moderate extension.

d. *Divergent luxation of the forearm;* the ulna being luxated backward, the radius forward, so that the humerus appears driven like a wedge between the ulna and radius. This injury is very rare. The several bony parts can be directly palpated in their abnormal position. During reduction every bone is to be treated by itself; the ulna by hyperextension and traction, then the radius by direct pressure.

e. *Isolated luxation of the ulna,* observed very rarely, results from a fall upon the hand while the forearm is hyperextended and pronated. Symptoms like those of a posterior luxation of the forearm, except that the capitulum of the radius is not found dislocated; the elbow is in varus position, the ulnar side of the forearm is shortened. Reduction by hyperextension and traction.

f. Isolated luxation of the radius, an injury which is less rare and occurs in different forms. The capitulum may be luxated forward, backward, or outward; the latter is very rare in a pure form, but is more often complicated with fracture of the ulna in the upper third; the capitulum can be felt at the outer margin of the external condyle; the radial side of the forearm is shortened and the elbow therefore is in valgus position. Reduction by direct pressure, if necessary by producing a varus position at the elbow.

The backward form is very rare, but easily recognized by palpation of the capitulum of the radius. The elbow is half pronated; active extension and supination are impossible. Reduction by direct pressure during vigorous traction and varus position of the forearm.

The forward variety is more frequent; resulting directly from a blow from behind against the capitulum of the radius, or from a fall upon the hand during pronation. The capitulum of the radius is situated in front above the eminentia capitata, and causes a prominence in the region of the supinators. The forearm is slightly flexed and pronated; active supination is impossible; flexion can be effected only to about a right angle. The radial side of the forearm is shortened unless fracture of the ulna in the upper third (which see) exists as an important complication. Reduction is best effected by vigorous traction, the elbow being bent and pronated.

In all these cases of isolated luxation of the radius the annular ligament is torn, or the capitulum has slipped out of it. Not rarely, especially in forward luxation, reduction is very difficult or impossible owing to interposition of portions of the capsule. In the latter case arthrotomy should be performed and the reduction forced after disengaging the interposed structures. A like operation is indicated in old cases, the method of operation being a radial longitudinal incision over the joint. If the incision were made from in front the radial nerve would be liable to be divided. Only in the most intractable cases would resection be indicated instead of arthrotomy.

From a practical point of view mention should here be made of an affection whose etiology and symptomatology are well known, but whose anatomical details are still in dispute. The affection occurs in small children and results from violent traction on the arm by the attendant, whether to avert a threatened fall or when the child is gliding from the lap, etc. The symptoms are: the child does not move the painful elbow and lets it hang down in a pronated position; there is no demonstrable deformity. Attempts at movements of supination are very painful, but supination with traction followed by flexion causes the pathological manifestations to disappear. The child then is able to move the arm again, but it is better to keep it at rest for a few days by means of a mitella. This symptom complex which always recurs in an extremely typical form is interpreted by some surgeons as the result of an incomplete forward luxation of the radius, by others as a consequence of an incarceration of the intact joint capsule (at its posterior surface) between the capitulum of the radius and the humerus.

Among incidental injuries lesion of the radial nerve has been particularly observed.

The after-treatment of all these luxations is carried out on general principles.

6. FOREARM.

The forearm is very frequently the seat of fractures; this fact finds its explanation in its functions during work and for protection, as it is stretched forward to ward off injuries. We distinguish fractures of the forearm, *i.e.*, of both bones, from isolated fractures of the ulna or radius.

A. Fracture of Both Forearm Bones.

(*Fractura Antibrachii.*)

This arises generally directly from a fall or a blow. In children infractions with curvatures of the forearm are not rare.

Symptoms.—Usually the displacement (*ad axin*) at once calls attention to the presence of the fracture; on careful examination abnormal mobility and crepitation are discovered. As the fractures as a rule occur in the median third of the forearm these signs are for the most part easily and positively demonstrated. Fractures of the forearm near the wrist will be discussed more fully in connection with the typical epiphyseal fracture of the radius. When both bones are fractured at the same level the displacement is as a rule greater than when the fracture of the two bones is at different levels. This fact, too, is of some importance for the prognosis. For a similar reason it should be noted whether the displacement has caused a lateral approximation of the two bones and extensive injury of the interosseous ligament. As such injury may be followed by cicatricial contraction and partial ossification of this ligament, and the bones may also come in mutual lateral contact, whether by osseous union or by a kind of conical articulation (Plate 38, Fig. 3), it is clear that the function of the forearm with reference to pronation and supination may be seriously impaired. While such complications are quite indifferent in the case of the leg, on the forearm they may cause permanent and grave interference with the capacity for work.

For this reason the treatment of fractures of the forearm is of special importance and should be carried out with care and skill. The aim is to secure osseous union of the fragments in good position of each bone, with unimpaired mobility of the two neighboring joints and the two bones on each other. It is furthermore essential not to do harm in a certain respect by the dressing; the latter may be well intended and fit snugly, yet it may cause direct injury if it presses the bones together laterally by circular turns, that is to say, if it approximates them at the point of fracture so that they may completely coalesce when the callus formation is abundant. Hence the splints should not be narrow but broad (perhaps made of pasteboard and strengthened by strips of wood), so as to project laterally some distance beyond the limb.

Another point which is of great importance even after the most careful reposition is the position given to the forearm in the dressing; of course the elbow is bent at a right angle, and the wrist-joint, extended, is included in the dressing.   But should the forearm or the hand be in pronation or in supination?   According to the preceding remarks, a position in which the ulna and radius cross is to be absolutely avoided; in this respect the parallel course of the two bones, that is almost complete supination, is the best.   Furthermore, the influence of muscular traction upon the fragments requires attention.   On Plate 39, Fig. 1, the influence of the biceps muscle on the upper fragment of the radius is illustrated; this muscle puts the bone in supination.

Therefore, were the dressing applied with the hand in pronation while the upper fragment of the radius remained in supination, the result would be a very faulty union, with subsequent loss of the movement of supination.

Every angular position of the radius at the point of its fracture is also liable to hinder the motions or the unfolding of the interosseous ligament, and thus limit the normal excursion with reference to supination.

Accordingly, after careful reposition of the fragments, the dressing, including a splint of sufficient width, should be applied in the supine position.   The splint may be applied to the dorsal or the volar side, best on both sides, using a longer and a shorter splint.   In these fractures in particular it is essential that the first dressing be well padded, not too firmly applied, and that the hand and fingers be frequently inspected, for it is in these very cases that gangrene and ischaemic muscular contractures have been observed as a result of tight dressing, especially a circular plaster-of-Paris bandage applied soon after the injury (see General Remarks). Moreover the change of the dressing after about a week and a careful examination of the position of the fragments at that time are very important. A threatening angular position salient on the extensor side may be successfully treated by an appropriate splint applied on the extensor side, the elbow being extended. During the second change of the dressing careful passive movements and massage are indicated. Irregularities in course, delayed callus formation, and the development of a false joint occur occasionally and are to be treated on general principles.

B. FRACTURES OF THE ULNA.

a. *Fracture of the Olecranon* (*Plates 30 and 37*).

This results commonly from a fall upon the elbow, that is, from a direct force, very rarely from muscular traction (by the triceps) or from hyperextension, the posterior surface of the humerus crowding against the process.

The symptoms are simple, as the fracture is nearly always transverse through the middle of the olecranon, with a distinct diastasis between the fragments; the upper fragment is drawn up by the triceps. As the olecranon has a superficial position it is readily felt. The joint and the remaining bony prominences in the articular region are intact; only the effusion of blood caused by the fracture is of course present also in the joint. Active extension of the flexed arm is impossible. Usually the upper fragment can be pushed down sufficiently to produce crepitation dur-

ing lateral movements. Should the fragments have
remained in contact (when the periosteal covering
and the lateral tendinous fibres are partly intact) the
prognosis is of course favorable, and the result will be
firm bony union. With diastasis of the fragments
osseous union is hardly to be expected; on the contrary, union is effected for the most part by connec-
tive tissue. This is to some extent due to the fact
that the fragments are bare of periosteum on the side
directed toward the joint and are covered with thick
cartilage, while on the outside is a tense layer of
fibres (attachment of the triceps tendon); as a result
the formation of callus is relatively scant.

Treatment.—The first aim has regard to the factors which cause the diastasis; the arm is to be
dressed in complete extension because thereby the
lower fragment is most closely approximated to the
upper, which is drawn up by the triceps. Furthermore it is sometimes useful to remove the effused
blood from the joint by aspiration if the quantity and
tension increase the diastasis of the fragments. Besides, the upper fragment should be brought as near
to the forearm as can be attained by manual fixation;
this is effected by one or more narrow strips of adhesive plaster which are looped around the tip of the
olecranon above and pass down on each side toward
the flexor side of the forearm. The primary bone suture of the fragments may be used under certain conditions, trusting to the aseptic success of this operation; however, it is not a method suitable for general
introduction but is dependent upon the favorable
auxiliaries presented in a clinic.

That the injury otherwise is to be treated as an articular fracture is obvious. It is important that massage of the triceps be begun early, and attention may be called to the fact that in most recent times the massage treatment of fractures of the olecranon, carried out in a manner similar to that in patellar fractures, has given good results.

b. Fracture of the Coronoid Process (Plate 37).

This injury is rare and is observed most frequently in combination with backward luxation of the fore-arm. Only when the coronoid process is broken off at its base is the separated fragment under the influence of the brachialis internus; for this muscle is not inserted at the point but some distance below it. The fracture in its pure form results particularly from a force which moves the lower end of the humerus toward the anterior side of the ulna, that is, toward the coronoid process.

The symptoms are those of a severe articular injury. Direct palpation of the fragments is impossible owing to the soft parts at the anterior side of the joint. Exact palpation shows that the bony prominences are intact, only the olecranon sometimes projects slightly backward (subluxation), but can be immediately replaced by traction on the forearm. With the elbow at an obtuse angle this displacement of the olecranon can be produced at once by backward pressure on the forearm and again replaced, when crepitation will be present.

The treatment requires complete reposition by forward traction on the forearm, followed by fixation in acute-angled flexion; the other steps are those of articular fractures in general.

c. Fracture of the Ulna in its Upper Third, with Luxation of the Capitulum of the Radius (Plate 36).

In the limbs containing two bones, the forearm and the leg, certain findings are typical and readily explained. If both bones are broken the fracture may be associated with more or less displacement; the condition of the one bone will resemble that of the other. If, however, only one bone is broken the other will act as a kind of splint and may undoubtedly prevent marked displacement. Therefore if we find a fracture of one bone with pronounced displacement of the fragments, the other must of necessity be broken likewise or have suffered some other displacement, a luxation. In practice the attentive physician will not fail to observe that fractures of the ulna, when marked displacement is present, are associated with luxation of the capitulum of the radius; fractures of the tibia, in like manner, with luxation of the capitulum of the fibula.

Fracture of the ulna in the upper third with considerable dislocatio ad axin and consequent shortening of the bone, associated with luxation of the capitulum of the radius (generally forward), is a typical injury. The illustrations on Plate 36 are quite characteristic and correspond completely with what I have repeatedly observed in the living patient. The evidences of the fracture are very distinct; there is never any difficulty in the diagnosis. On the other hand, the injury at the elbow joint, the luxation of the radius, is often overlooked. Any one paying attention to the introductory remarks will not commit this error. The displacement of the fragments is so marked, the consequent shortening of the ulna in its longitudinal direction is so great, that the radius must of necessity be likewise fractured or luxated. The surgeon who examines the elbow-joint will miss the capitulum of the radius at its normal site and will find it in luxated position at the external condyle or on the anterior side of the joint. The prognosis is favorable if the correct diagnosis is made early; for the reposition as a rule offers no particular difficulty when performed under anæsthesia. Vigorous traction on the forearm must obtain the correction of the fractured position, while direct pressure with a view to replacement is exerted on the head of the radius during flexion of the forearm. The head of the radius sometimes shows a tendency to renewed luxation or forward subluxation; for this reason the dressing is properly so applied, the forearm in supine position being flexed at least at a right angle, that a soft pad in the bend of the elbow will exert gentle pressure on the capitulum of the radius.

In old cases of this nature osteotomy at the point of fracture and arthrotomy for the reposition of the head of the radius or its resection are required.

d. Fracture of the Diaphysis of the Ulna.

When a person falling puts forth the arm so as to strike on the forearm with the elbow bent, or tries to ward off a blow with the arm, the impact is received chiefly by the ulna, which may be fractured. These are direct fractures, and they may with justice be called parrying fractures. Such injuries result very rarely from indirect force. The diagnosis is

easily made, since, owing to the superficial position
of the bone, abnormal mobility and crepitation are
positively demonstrated. The treatment is the same
as for fractures of both forearm bones; marked displacements are hardly liable to occur where the radius is intact.

e. *Fracture of the Styloid Process of the Ulna.*

This occurs very rarely in an isolated form, when
it may be demonstrated by careful palpation. A
pseudarthrosis is very apt to develop during the healing process.

Further details about this fracture will be found in
the section on typical fracture of the lower epiphysis
of the radius.

C. FRACTURES OF THE RADIUS.

a. *Fracture of the Capitulum of the Radius (Plate 31, Fig. 4).*

This of course presents the symptoms of an articular lesion and undoubtedly is not rarely mistaken for
a simple contusion or distorsion of the joint. The
fracture is altogether intra-articular; it may be complete or incomplete (fissure or infraction). In the
latter case the diagnosis is obviously difficult and
doubtful. Complete fractures are recognized by the
fact that the capitulum is separately and abnormally
movable under crepitation, though this is not always
the case. Particularly during pronation and supination the associated movement of the capitulum is
often undisturbed. The pain, of course, is limited
to the region of the capitulum.

The fracture results sometimes directly, more frequently indirectly from a fall upon the hand with the elbow extended or flexed, when a marginal portion of the capitulum is forced off against the eminentia capitata (so-called chisel fracture).

Treatment.—As we cannot act directly upon the separated fragment, union is often effected with considerable displacement. Evidently the indication is to apply a dressing which puts the elbow and wrist joints at rest, and perhaps exerts direct pressure upon the region of the capitulum of the radius. Nevertheless, despite the employment of the auxiliaries indicated in articular fractures, considerable stiffness of the elbow-joint not rarely is left behind and subsequently calls for resection of the capitulum.

Lesion of the radial nerve has been occasionally observed as an incidental injury.

Fractures of the neck of the radius, that is to say, below the capitulum, are very rare. When present the capitulum does not move with pronation and supination of the hand; a bony prominence may also be felt at the seat of the injury. Treatment as above.

Traumatic separation of the epiphysis at the upper end of the radius is very rare, and of course occurs only in children.

b. Fracture of the Diaphysis of the Radius.

Though fractures of the shaft of the ulna are frequent, those of the shaft of the radius are rare. They may have a direct and an indirect causation. As the symptoms are distinct, the diagnosis is readily made. As regards the displacement and the treatment, see the section on fractures of the forearm.

c. Fracture of the Lower Epiphysis of the Radius (Plates 40, 41, 42).

This fracture is very frequent and practically of the greatest importance; it is justly called typical, because its symptoms are extremely characteristic and despite minor differences come again under observation in every case of the kind.

This typical fracture of the radius belongs to the group of supracondylar fractures, that is to say, the line of fracture is usually located about one-half to two centimetres above the lower articular surface, in other words at the point where the compact bone of the diaphysis passes into the markedly cancellous expansion of the articular extremity; at the limit of these two portions fracture is more liable to occur for anatomical and mechanical reasons. Sometimes, however, the lower fragment does not include the entire articular extremity; the line of fracture may also pass through the epiphysis proper and lead merely to the separation of a smaller fragment.

The cause of the fracture is almost invariably a fall upon the volar side of the hand. The first result is a hyperextension (dorsal flexion) which is checked by the strong mass of ligaments at the flexor side of the wrist joint (ligamentum carpi volare); greater force and continuation of the movement, however, do not cause laceration of this ligament, but, through its influence upon the lower end of the radius, fracture at the point named. This explanation, therefore, proves this to be a pronounced fracture by traction, and this view is generally accepted. Only in most recent times has this theory been disputed and the claim made that during dorsal flexion of the hand the upper row of carpal bones is forced against the dorsal prominence of the lower end of the radius, so that the fracture would be due to inflexion rather than traction. But whether the fracture be the result of traction or of inflexion, the acting force always causes at the same time a dorsal displacement of the separated lower fragment. That the lower end of the ulna as a rule does not suffer injury is easily understood from the anatomical arrangement of the parts, since it has no direct connection with the wrist-joint itself.

When the fracture results from a fall upon the dorsum of the hand, which has been observed, though rarely, the peripheral fragment generally is not displaced dorsally but toward the palm.

The symptoms of this fracture must be determined by careful examination, a minute inspection being the first step. It is best for the physician to sit directly opposite the patient, whose forearms are bared and who places his two hands side by side in symmetrical position. If a fracture is present, inspection shows the following: the region of the injured wrist-joint is altered, the styloid process of the ulna projecting more strongly than on the healthy side (compare Plate 40, Figs. 1 and 2). The hand near the wrist-joint is displaced radially; but on tracing the longitudinal axis in the middle of the forearm on each side we find that this line on the healthy side strikes about the centre of the middle finger, while on the injured side it passes more toward the ulnar side. The region of the styloid process appears widened. All these symptoms are due to the fact

that the peripheral fragment (the separated epiphysis
of the radius) is displaced radially.

Then follows inspection from the side, best the radial. In a healthy arm the lower end of the pronated forearm presents at the radius a somewhat
sinuous line, convex toward the dorsum and concave
toward the palm. On the fractured arm this line is
changed, usually in the opposite way: there is an
abnormal protrusion on the flexor side and a slightly
depressed angle on the dorsal side. On tracing the
longitudinal axis of the forearm, say with a blue
pencil on the skin, this line on the healthy side passes
straight over the region of the wrist-joint. It is interrupted, however, on the injured side, the line being
bent in above, corresponding to the epiphysis of the
radius; in this way, when the hand is extended
straight, a bayonet-like direction results which is
characteristic of this fracture. This kind of displacement finds its simplest explanation by the continuation of the force acting during the injury. As
soon as the fracture has resulted, the weight of the
falling body must continue to act until the diaphyseal end of the radius reaches the ground. In this
way the epiphyseal fragment suffers an upward displacement; it comes, as it were, in a somewhat supine position, while the shaft of the radius undergoes
pronation. In this occurrence, of course, the connection of the lower end of the radius with the ulna is
of importance: the displacement is effected in such a
way that the lower end of the ulna forms approximately the centre for the movement of the radius in
consequence of the ligamentous connection between
the ends of the two bones. It is possible that muscu-

lar action may play a part in the production of this typical displacement, but the main point lies in the force at work, as has just been briefly explained.

The other symptoms of a fracture are not always pronounced. Abnormal mobility is generally not easily demonstrable; in order to prove its presence the epiphyseal fragment must be very firmly fixed and the injured arm be given some support by resting it against the examiner's body. But it is not necessary to force the demonstration of this symptom. A similar remark applies to crepitation, but a characteristic snapping or rubbing is more frequently felt. Of greater importance is the demonstration of the painful spot; on palpating the radial side of the articular region the line of the joint itself and even the styloid process of the radius will be painless, while about one or two centimetres above the typical. pain is experienced. During this palpation the result of the inspection is confirmed. We feel particularly the abnormal bony prominence at the seat of the fracture on the volar side and the depressed angle on the dorsal side.

The prognosis of the fracture depends in the main upon the treatment.

Treatment.—The first requirement is to effect exact reposition, which is done by forced flexion and traction, best under anæsthesia. In many cases, after careful reposition, there is no tendency for the displacement to recur. Still it is desirable to follow certain rules in applying the dressings. The latter should include the entire forearm, the wrist-joint, and the metacarpals. The elbow-joint need not be, and the fingers must not be, fixed; for in many patients the fixation of the fingers is very apt to lead to excessive stiffness, which subsequently requires painful treatment (massage and mobilization) and sometimes cannot be completely relieved.

In order to retain the lower fragment in place the hand should be in a certain position, this being the only way in which the short fragment can be acted upon. The hand must be in volar and at the same time in ulnar flexion; thus the recurrence of the displacement will be prevented. Reposition should not be forgotten, nor during the dressing that the hand (with the fragment) as a whole is displaced toward the ulna, otherwise a disfiguring prominence of the styloid process of the ulna will remain behind.

It is immaterial whether this object is attained in one or the other way when the dressing is applied. A Beely's plaster-of-Paris splint (Plate 42, Fig. 2) is quite suitable, or else the application of a small bent splint which fixes the hand in the desired position. If a splint must be improvised from a piece of pasteboard or wooden board, only ulnar flexion can be secured by the form of the splint (pistol-shaped splint); in that case it will be a good plan to place a soft roller under the epiphyseal end of the radius so as to keep it somewhat elevated; while the diaphyseal end, being not so supported, sinks a little. Roser's splint dressing (Plate 42, Fig. 3) is applied in full supination, the patient as it were looking into his own palm. This dressing is somewhat bulky but gives good results. Of course it is essential that the injury be treated as an articular fracture, with frequent change of dressings, early massage, etc. Quite recently attention has again been called to the fact that the fracture, if the fragments have been properly replaced, will unite without any dressing, being simply placed in a mitella, with the best results as regards mobility of the wrist-joint. But for various reasons this is not to be recommended for general acceptance. There is no question that it is always better to have the fracture unite with some displacement but with good mobility, than without displacement and with great impairment of the function of the wrist-joint.

When the fracture is associated with a fracture of the styloid process of the ulna, that is to say, in the case of a fracture of both forearm bones at their lower ends, the wrist-joint is particularly liable to be involved. In general this fracture is to be treated on the same principles; in some cases it has been found necessary later on to resect the styloid process of the ulna in order to improve the mobility.

D. LUXATION IN THE LOWER RADIO-ULNAR ARTICULATION.

Despite the weak ligamentous apparatus, and despite the frequency with which this region is acted upon by extraneous forces, this luxation is very rare. The lower articular end of the ulna may be dislocated dorsally (directly by a fall or through excessive pronation), or toward the palm (directly or through excessive supination). The symptoms are determined by careful palpation. A subluxation at this joint occurs in washerwomen from wringing clothes. The treatment is carried out on general principles.

7. WRIST-JOINT.

Luxation of the hand at the radio-carpal joint is extremely rare. Although this diagnosis was formerly often made, it is now known that the large majority of cases were typical fractures of the epiphysis of the radius. The cases of true luxation which have been positively demonstrated may easily be counted (about thirty), and moreover are complicated in part with fracture of the styloid process of the radius.

The luxation may be dorsal or volar; the carpus then is situated on the dorsal or volar side of the articular ends of the forearm bones. The injury results from a fall upon the outstretched hand in pronounced dorsal (dorsal form) or volar (volar luxation) flexion. The diagnosis is made by careful palpation; reduction is effected by traction and direct pressure.

8. HAND AND FINGERS.

A. FRACTURES.

Fractures of the carpal bones are rare and have been observed usually in combination with severe lacerations or contusions of the overlying soft parts. The grade of the lesion depends upon these complicating injuries.

Fractures of the metacarpal bones are not so rare and result directly from impacts or blows upon the dorsum of the hand. Usually abnormal mobility and crepitation can be demonstrated, together with violent pain at the point of fracture. Displacement is absent as a rule, for the bones are, as it were, splinted by one another. The treatment therefore is simple. Unless direct pressure be required occasionally for the retention of a fragment, a simple circular bandage and mitella suffice; early massage and exercise of the fingers are useful.

Fractures of the phalanges, usually from direct force, may also result indirectly from a force acting in the longitudinal direction of the phalanges (with consequent longitudinal fracture); it is said that they may occur on the ungual phalanx even from traction of the extensor tendon during forced flexion (fracture by traction). The diagnosis and treatment (small padded splint) of these injuries are very simple, owing to the superficial location of the parts.

B. Luxations.

Luxation at the intercarpal joint, the two rows of carpal bones being displaced on each other, is extremely rare. The luxation of single carpal bones is not quite so rare. Of course the dislocated bone forms a prominence by whose shape and location the diagnosis is made.

Luxation at the carpo-metacarpal joints likewise has been rarely observed, the most frequent being the one at the thumb, where a dorsal, more rarely a volar, and a radial dislocation of the first metacarpal occur. The abnormal prominence and the direction of the shaft of the metacarpal determine the diagnosis; reduction is effected by traction and direct pressure.

Luxation at the metacarpo-phalangeal joints (Plates 43 and 44) is rare on the four fingers, but is more frequent on the thumb, and very important prac-

tically. Luxation of the thumb as a typical injury is
always dorsal, that is to say, the base of the first
phalanx is dislocated upon the dorsal side over the
head of the first metacarpal. The luxation of the
thumb may be incomplete, when the two articular
surfaces are still in contact, or complete, when they
are fully separated. This luxation can be easily produced artificially in the cadaver by hyperextension
(maximal dorsal flexion) and a vigorous backward
push toward the wrist-joint of the first phalanx thus
dislocated. When the thumb by some slight flexion
is again brought into an almost straight position, all
the characteristic symptoms of the typical luxation of
the thumb are present. I have even observed interposition with impossible reduction, which will be discussed hereafter.

The essential point in artificial luxation, as in that
observed during life, is the fixation of the dislocated
thumb. This results from the traction of the soft
parts appertaining to the joint and surrounding it.
The lateral ligaments frequently are not torn, and a
number of powerful muscles and tendons effect the
fixation by closely hugging the head of the metacarpal at the same time. Thus is caused a kind of obstruction to the reduction when the faulty effort is
made to do so by simple traction; the stronger the
traction the more closely and tightly the tendons and
muscles hug the neck of the capitulum, and thereby
increase the difficulty (the so-called buttonhole mechanism, compare Plates 43 and 44, Fig. 1).

Symptoms.—The bayonet-like direction of the
thumb in connection with the first metacarpal, together with the marked prominence of the head of

the latter bone on the volar side, the presence of the abnormal direction of the first phalanx, as. well as the peculiar firm fixation in this position—all these lead to the correct diagnosis.

The reduction, as in all hinge joints, should be effected without force. The thumb is first hyperextended and then pushed forward by direct pressure against the base of the first phalanx. As soon as the greater part of the articular surfaces have come into normal contact flexion will succeed and the reduction is complete.

Stress is to be laid upon the correct performance of this manipulation, and yet the reduction may fail in spite of it.

A frequent cause of obstruction is found in the interposition of the capsule, sometimes also of the sesamoid bones. In other cases I have found a peculiar relation of the tendon of the flexor pollicis longus to be a cause of obstruction, and this not only during life but also when the luxation was artificially produced in the cadaver. The tendon surrounds the neck of the first metacarpal; when, as sometimes happens, the articular surface of the capitulum terminates on the ulnar side in a thick expansion, the tendon may be almost incarcerated behind it, and reduction be rendered impossible. This condition may sometimes be recognized by a slight inclination and twisting of the thumb toward the ulnar side; occasionally but not always the incarceration of the tendon may be overcome by a greater inclination toward this side.

Exceptionally it may happen during reduction, especially if carried out incorrectly by traction upon the thumb, that the capsule and the external sesamoid bone are turned over and interposed in an inverse direction between the articular ends (complex luxation).

Reduction having failed, arthrotomy should be immediately performed. In all cases of this nature I have succeeded, by a preparatory incision over the capitulum projecting on the volar side, in finding the obstruction to the reduction, in effecting the latter, and in securing a movable joint. In very old cases resection of the capitulum might be necessary.

After these explanations and a careful inspection of Plates 43 and 44 nothing further need be said regarding the symptoms and treatment of the much rarer volar luxation of the thumb and the somewhat more frequent luxation at the interphalangeal joints. The above-described relations of the dorsal luxations of the thumb apply more or less closely also to these forms of dislocation.

V.

FRACTURES AND LUXATIONS

OF THE

LOWER EXTREMITY.

FRACTURES OF THE PELVIS. BERTINI'S LIGAMENT.

FIG. 1.—*The lines of fracture in the anterior circumference of the pelvic ring,* on both sides of the symphysis pubis, correspond to those which may be produced artificially by compressing the pelvis in a direction from before backward. The fracture shown resulted from being run over; the patient, an adult, lying on the back. Besides the separation of the anterior middle portion there is a *diastasis of one sacro-iliac symphysis.* In other cases there is in addition a fracture of the sacrum or of the venter of the ilium. Sometimes there occurs in one half of the pelvis a fracture in front through the bones surrounding the obturator foramen, behind through the venter of the ilium, *i.e., Malgaigne's double vertical fracture.*

FIG. 2.—*Pelvic fracture through the acetabulum* in a boy aged 14 (W. Kohn, 1889; see description of Fig. 1, Plate 1). He was injured by the cam-wheels of a threshing machine. Among the incidental injuries was a large flap wound of the left inguinal region, at the bottom of which the femoral vessels lay as if dissected free, and which led into a large wound cavity between the adductors, where the bones surrounding the obturator foramen could be felt to be fractured. The left thigh was slightly adducted and appeared shortened. The penis was completely flayed. The urethra was intact; the catheter evacuated normal urine. The wounds were dressed without anæsthesia and the bleeding vessels ligated; infusion of 300 cc. salt-and-sugar solution. Still, collapse and death ensued in a few hours. The fracture involved the left os pubis and the ischium, and had caused in the acetabulum a wide diastasis of the Y-shaped symphysis.

FIG. 3.—Illustration of the *ileo-femoral or Bertini's ligament,* whose importance for luxations of the hip-joint was demonstrated by Bigelow. The femur is in the position of an iliac luxation.

54

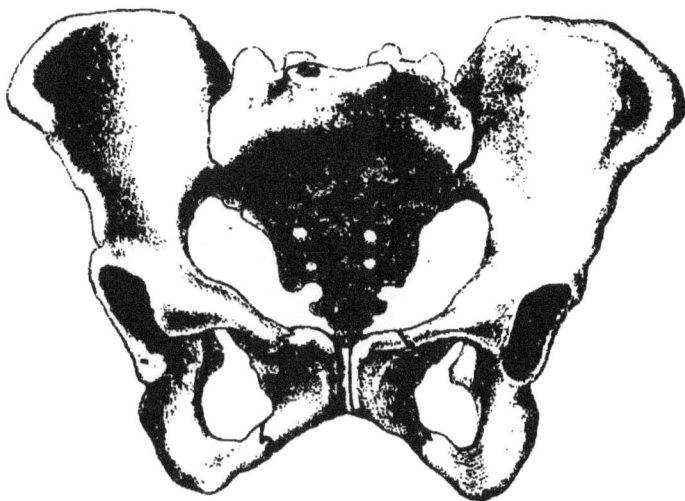

Fig. 1

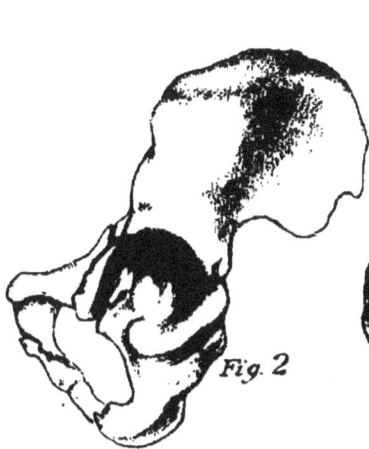

Fig. 2

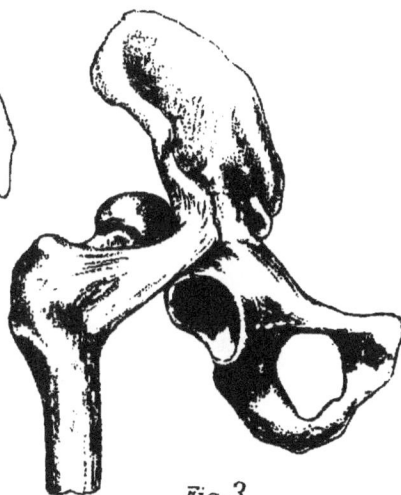

Fig 3

Lith Anst v. F. Reichhold, München

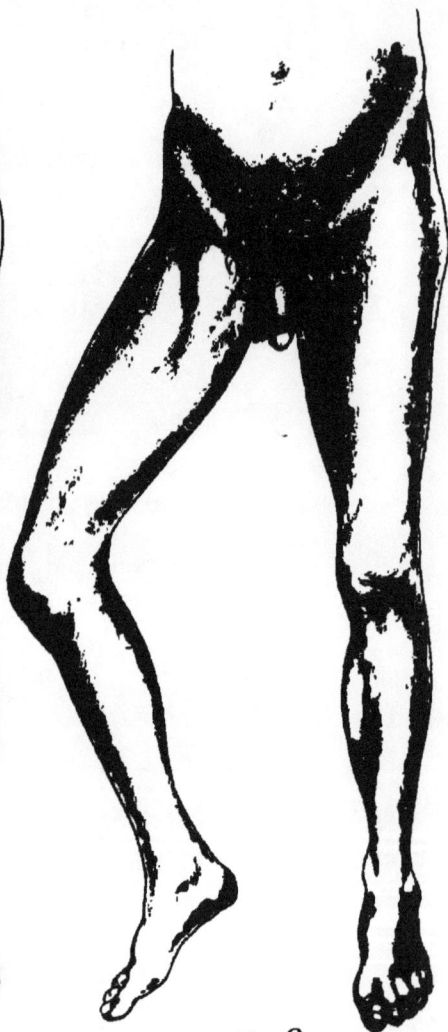

Fig. 1

Fig 2

Lith. Anst. v. F. Reichhold, München

Explanation of Plate 46.

LUXATION OF THE THIGH.

The normal rotary movements at the hip-joint are inward and outward. On looking at a person from in front and bearing in mind the normal form of the upper end of the femur, it is easy to conceive how these rotary movements are effected around the head of the femur as a centre and with the length of the neck of the femur as the radius. When outward rotation is forced the joint capsule at its anterior circumference is rendered very tense, and when this outward rotation is continued the capsule tears in front and the head of the femur may undergo forward luxation. When inward rotation is forced the joint capsule may tear in its posterior circumference, and a posterior or backward luxation of the femur may result. Both forms of luxation occur more readily when the thigh is bent at the same time to about a right angle in the hip-joint, but even then powerful force is required, and not rarely an injury of the ligamentous apparatus or of the bones occurs previously at the knee-joint, which renders an increase of the movement leading to luxation impossible.

FIG. 1 shows a *backward luxation of the femur;* we see the characteristic inward rotation of the thigh with moderate adduction. The illustration was drawn from a photograph of an adult whose iliac luxation was reduced without difficulty under anæsthesia.

FIG. 2.—*Forward luxation of the femur;* the head is in the region of the obturator foramen. The thigh is rotated outward and flexed farther. This obturator luxation was produced artificially and the drawing made from a photograph.

LUXATION OF THE THIGH.

Posterior luxation of the femur. This luxation was artificially produced in the cadaver, and then used for the preparation here illustrated.

On the plate the normal head of the femur can be recognized at once; it is dislocated backward and could be felt there through the overlying soft parts. A portion of the neck of the femur is also visible. The glutæus maximus is divided in the direction of its fibres and the two portions are drawn wide apart. Beneath the upper portion of the glutæus maximus, which is fixed by a hook, a strip of the glutæus minimus is visible, and beneath the latter the pyriformis; we see it emerging from the lesser pelvis (whose margin appears above) and passing to the great trochanter immediately over the luxated head. In a median direction from the head of the femur the sciatic nerve, distinguished by a yellow color, is dissected free; it is also easily recognizable by its course. Between this nerve and the glutæus maximus we see the tuberosity of the ischium, and the tendon of the biceps femoris springing from it. Under the head of the femur some other muscles are visible which surround the femur like a cravat. These are, first, above the obturator externus and beneath it the quadratus femoris, whose fibres are partly lacerated. Between the middle of the head of the femur and the sciatic nerve a reddish strip of muscle is to be seen; this is the obturator internus, which is situated between the head and the margin of the acetabulum. Hence, after what has been stated, this is the so-called sciatic luxation. The illustration is very characteristic and gives a good idea of the large mass of muscles whose tension is to be overcome in the reduction of luxations at the hip-joint. For this reason profound anæsthesia is always indicated, in order that the reduction may be accomplished when the muscles are relaxed.

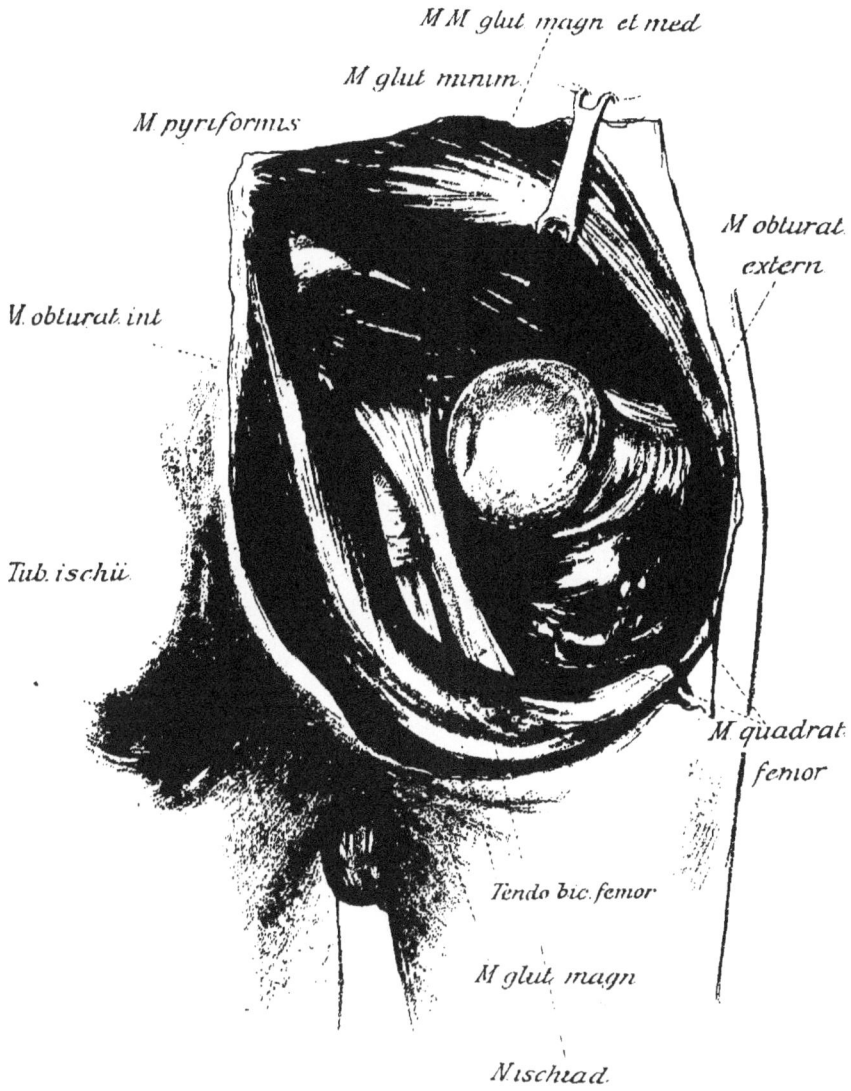

MM glut magn et med

M glut minim

M pyriformis

M oblurat extern

M. oblurat int

Tub. ischii

M quadrat femor

Tendo bic femor

M glut magn

N ischiad.

Lith Anst v F Reichhold, München

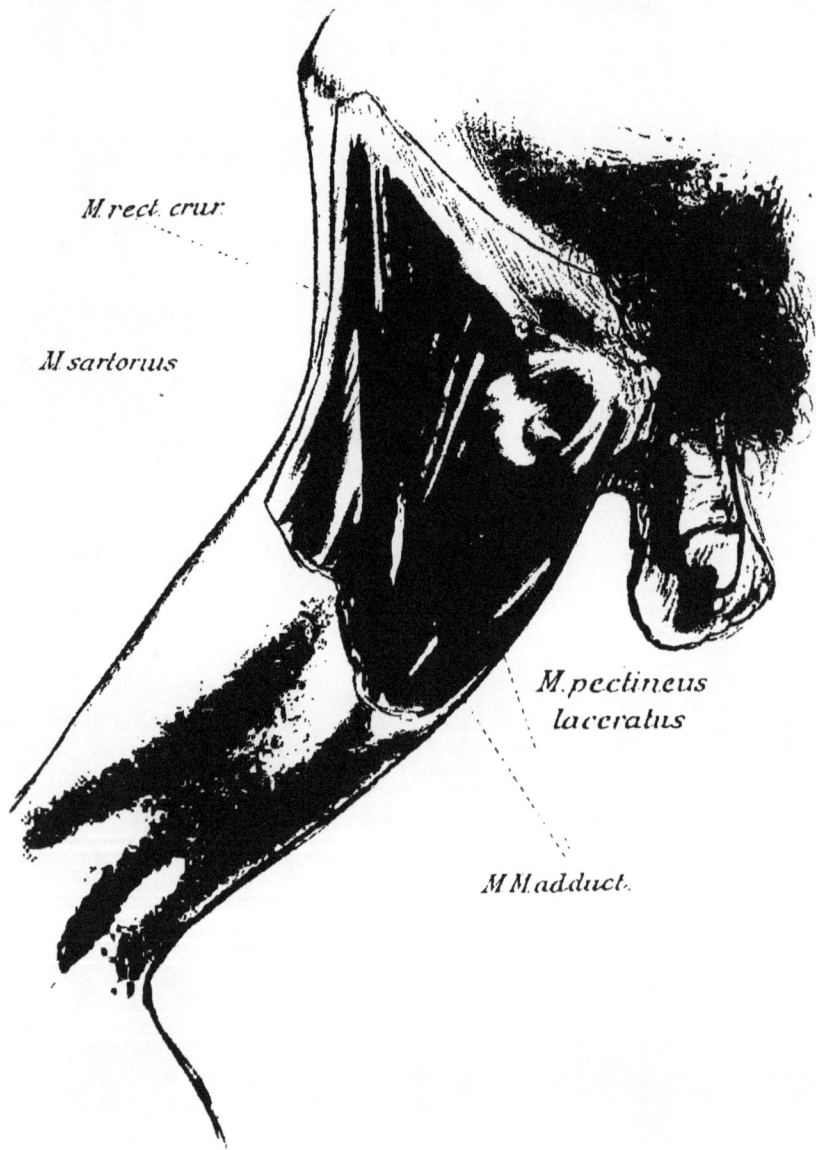

M. rect. crur.

M. sartorius

M. pectineus
laceratus

M. M. adduct.

Explanation of Plate 48.

Forward luxation of the femur. This luxation likewise was artificially produced in the cadaver and then dissected and drawn.

The position of the entire limb is characteristic. Here, too, we recognize at once the head of the femur covered by the torn fibres of the pectineus muscle. This laceration of muscles occurs also in the living subject, and for this reason the effusion of blood in luxation of the hip is often very great; but in this instance it may also be in part a post-mortem phenomenon, that is to say, it may depend on the greater friability and brittleness of the muscles in a cadaver which is not quite fresh. Inward from the head of the femur are the adductors; outward and upward are the femoral vein and artery, then the crural nerve (yellow) imbedded in the muscle, then the rectus femoris and the sartorius. Between the femoral vein and the head, above the latter, is the region of the crural ring.

It may be easily recognized that the head of the femur is situated in the region of the obturator foramen; it is an *obturator luxation.* Reduction was readily effected by inward rotation of the slightly flexed thigh.

A second variety of forward dislocation, as is well known, is the *pubic luxation*, in which the head is situated on the crest of the os pubis; the thigh is likewise rotated outward and flexed, but not nearly so much as in obturator luxation.

57

Explanation of Plate 49.

EXTRACAPSULAR FRACTURES OF THE NECK OF THE FEMUR.

It is a well-known fact that *fractures of the neck of the femur* are divided into extracapsular and intracapsular, according to whether the line of fracture separates the neck nearer to the great trochanter (extracapsular) or closer to the head (intracapsular). These terms are not quite exact as expressing the relation to the joint capsule, but they may be retained in the above anatomical sense.

All the illustrations on Plate 49 represent *extracapsular fractures* of the neck of the femur.

FIG. 1 *a* and *b.*—Fracture of the neck of the femur of an old woman who was injured by a fall upon the trochanter. The fracture is a pronounced extracapsular one, it passes even within the trochanter and forms the condition of the fragments known as *impaction* (gomphosis). In Fig. 1 *b* we see the external form, in Fig. 1 *a* the section of the specimen. The impaction manifests itself in the shortening of the neck of the femur, which, moreover, is almost at a right angle to the shaft of the femur; on the great trochanter we also recognize the infracted form. The *red line* drawn on Fig. 1 *a* shows the form of the *upper end of the femur* of the healthy side, divided in like manner. This marks very clearly the shortening of the fractured bone. (Author's collection.)

FIG. 2 *a* and *b.—United extracapsular fracture of the neck of the femur* in a woman (Glöwe) aged 82, who was injured in November, 1888. After her death (in the beginning of March, 1893), the specimen shown was found at the autopsy. It was impossible to make a comparison with the opposite side because it also contained a fracture of the neck of the femur, which is illustrated in Fig. 2, Plate 50. (Author's collection.)

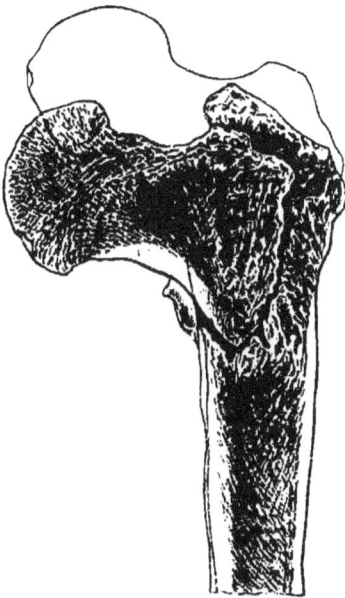

Fig. 1ᵃ

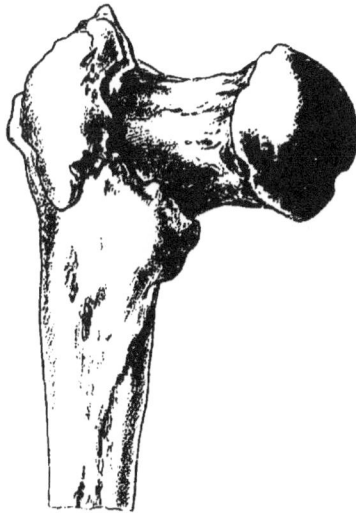

Fig. 1ᵇ

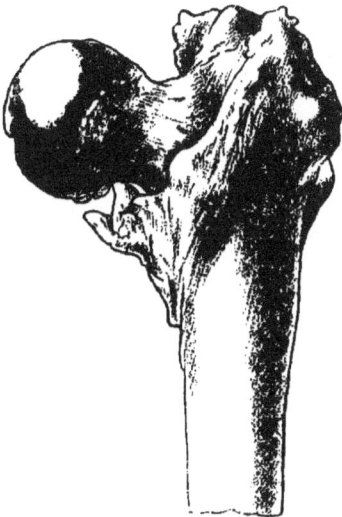

Fig. 2ᵃ

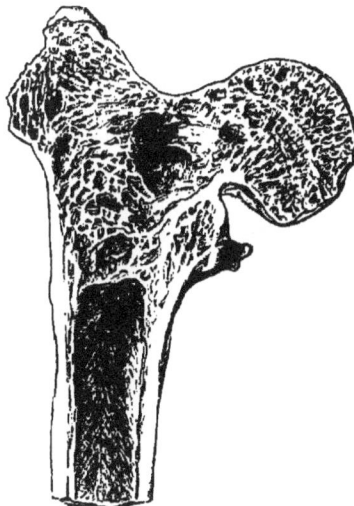

Fig. 2ᵇ

Lith. Anst .v. F. Reichhold, München.

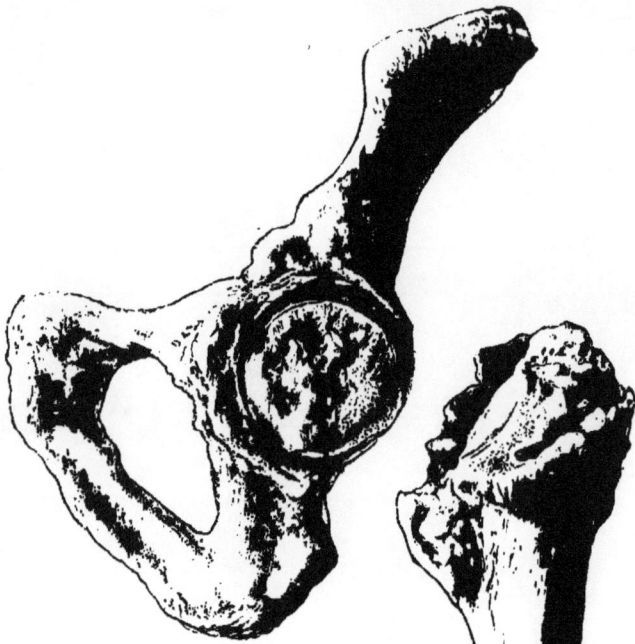

Fig. 1ᵃ

Fig. 1ᵇ

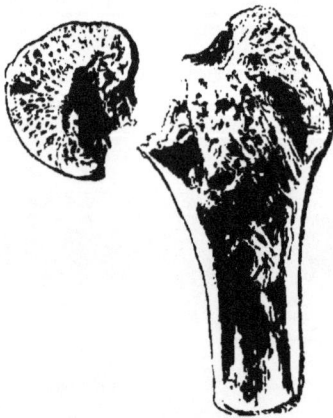

Fig. 2ᵃ

Fig. 2ᵇ

Explanation of Plate 50.

We see very distinctly that the neck of the femur has retained its connection with the trochanter and the shaft, and that really only the head of the femur is broken off.

The specimen illustrated in Fig. 1 *a* and *b* is extremely characteristic of those cases in which the head of the femur is situated in the acetabulum and its fractured surface corresponds with the level of the margin of the socket. Between this fractured surface of the head and the fragment of the shaft a kind of nearthrosis has developed in consequence of the displacement there occurring. The neck of the femur has been worn away in the course of time so that it has actually disappeared. Of interest is a condition of pronounced arthritis deformans which has led to deposits of bone at the margin of the acetabulum and the upper end of the shaft of the femur; the latter is thickened almost to a club shape by massive bone formation, and is flattened at its articulation with the pelvis, where it shows in part a dense spongy mass of bone, in part also distinct abraded surfaces such as are frequent in arthritis deformans; at the fractured surface of the head similar abrasions are barely indicated. (Author's collection.)

FIG. 2 *a* and *b*.—This is the specimen from the opposite side of the fracture of the neck of the femur shown in Fig. 2, Plate 39, from the same woman, aged 82, namely, *an intracapsular fracture of the neck of the femur with impaction.* The shortening of the neck of the femur caused thereby is especially marked in Fig. 2 *b*. Here, therefore, the fractured end of the neck of the femur is wedged in the cancellous tissue of the head, in the same way as it may be impacted in the mass of the trochanter in extracapsular fractures.

Explanation of Plate 51.

FIG. 1.—Picture of Ernst Gottschalk, a boy aged 8, admitted October 30th, 1889, with a *badly united fracture of the femur*. His thigh had been fractured on March 23d of the same year, and had been treated with plaster-of-Paris dressings. The deformity on admission is apparent in the illustration drawn from a photograph; the right thigh is considerably shorter than the left, hence the right half of the pelvis is lowered; besides the thigh shows an *angle salient forward and outward*. The gait of course was limping and laborious. Under anæsthesia osteoclasis at the point of fracture succeeded without difficulty. By the aid of a correct adhesive-plaster traction dressing and a pull by heavy weights, the thigh being in moderate flexion and abduction, union was secured in a straight line with barely any shortening.

The deformity alluded to is typical in fractures of the thigh at or slightly above its middle. It is due to the ileo-psoas muscle acting unilaterally upon the upper fragment and to the muscles inserted at the great trochanter (glutæus maximus, etc.).

FIG. 2 is to call to mind these muscles and their action upon the upper fragment of the femur. The illustration is carefully drawn from nature (preparation). Of the trochanteric muscles only the glutæus medius could be represented, because it alone reaches far enough forward on the crest of the pelvis to be seen in this view.

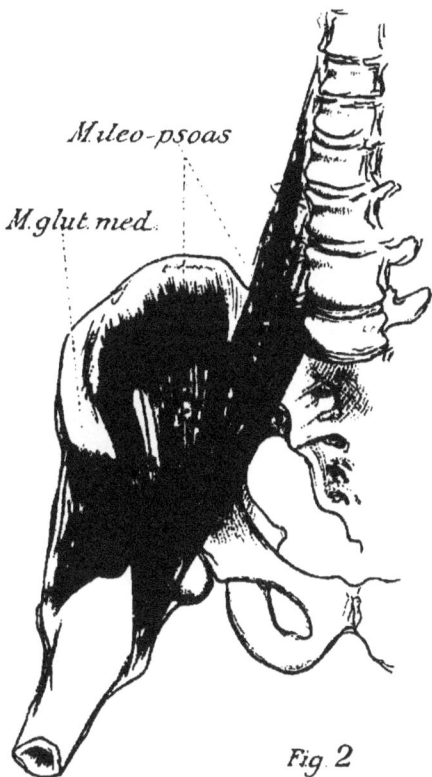

M.ileo-psoas

M.glut.med.

Fig. 2

Fig. 1

Lith.Anst.v.F.Reichhold,München

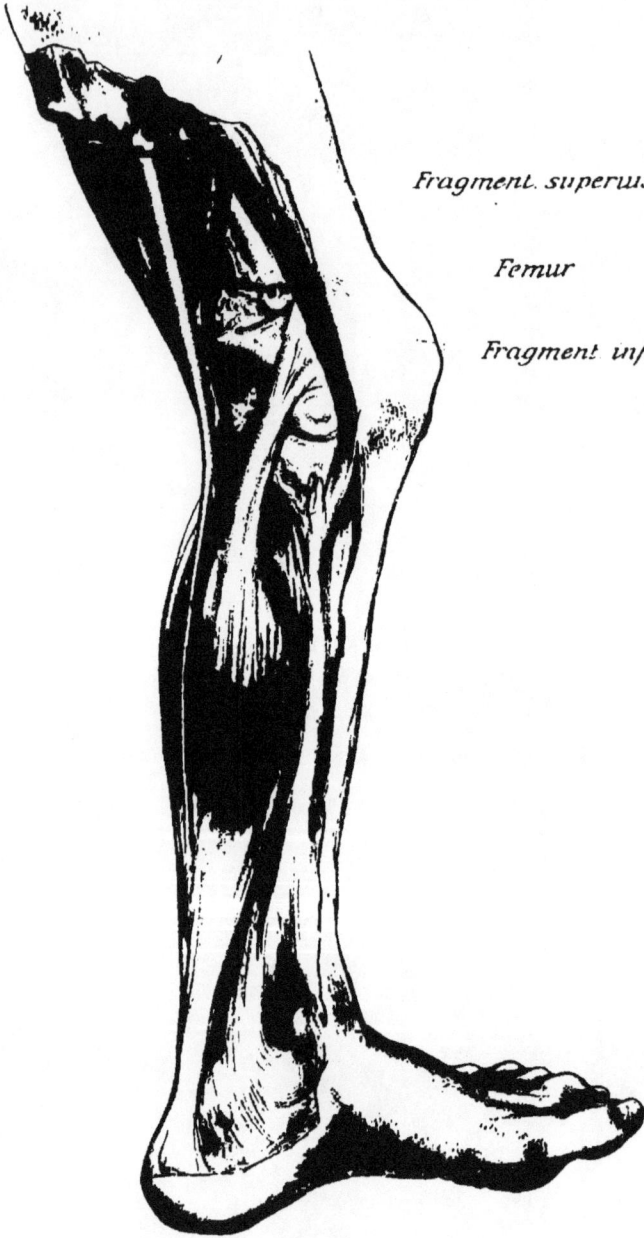

Fragment. superius

Femur

Fragment. inferius

Explanation of Plate 52.

The illustration was drawn from an artificial preparation in which the displacement of the fragments
was made analogous to the typical displacement in the
living patient. In the illustration we notice at once
the fractured surface of the lower fragment which is
placed in flexion by the traction of the calf muscles
and therefore projects backward. Above it may be
seen the shaft of the femur, the upper fragment.
Very interesting is the relation of the vessels (only
the artery is shown here), which as it were ride upon
the protruding edge of the lower fragment—a condition which has been repeatedly noted in the litera-
ture and which sometimes has led to gangrene of the
extremity.

A similar displacement has also been observed in
traumatic separation of the epiphysis at this point,
in young persons.

In order to prevent a displacement of the fragments
in this way we must, after careful reposition, resort
to a traction dressing with heavy weights and to
direct pressure against the lower fragment close above
the popliteal space by pulleys or wound rollers. Occasionally, however, it may be necessary to bend the
knee at a right angle and apply the dressing so as to
secure forward traction, in a manner resembling the
older treatment of fracture of the femur by the so-
called equilibrium method or the double inclined
plane, etc.

61

Explanation of Plate 53.

Different Fractures of the Femur.

Fig. 1.—*Very acute-angled oblique fracture in the upper half of the femur.* The line of fracture extended above into the great trochanter. The fracture has united without displacement and an abundant external callus surrounds the fractured surfaces of both fragments. (Pathologico-Anatomical Institute in Munich.)

Fig. 2.—*Oblique fracture below the middle of the femur* (right femur in anterior view). The fragments have united with slight displacement. We observe the external callus connecting the fragments in broad layers, and recognize on the surface of the upper fragment rarefying processes which in the course of time may bring about the definitive form of the bone corresponding to its mechanical efficiency. (Author's collection.)

Fig. 3.—*Old fracture of the femur united with marked displacement.* (Pathologico-Anatomical Institute in Munich.)

Fig. 4.—*Oblique fracture through the lower articular end of the femur,* with separation of the whole internal condyle of the femur. This injury (separation of one of the condyles of the femur) may cause severe lesions of the joint, the development of a genu varum or valgum, as well as arthritis deformans. In some cases we find at the lower end of the femur a transverse fracture combined with a longitudinal fracture extending into the joint, *i.e.,* the so-called T-fracture.

62

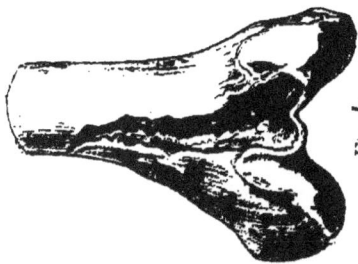

Fig. 4

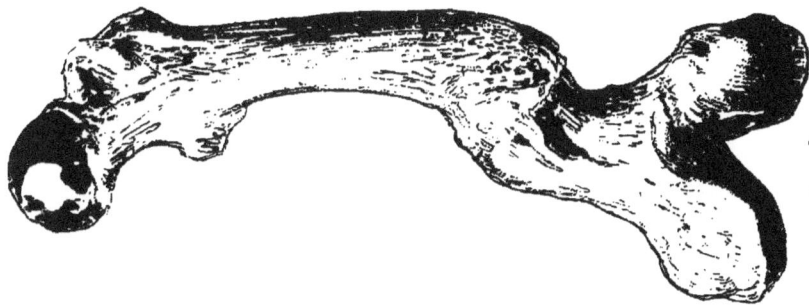

Fig 3

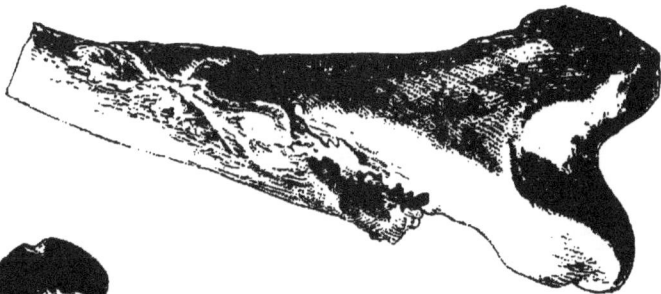

Fig 2

Fig. 1

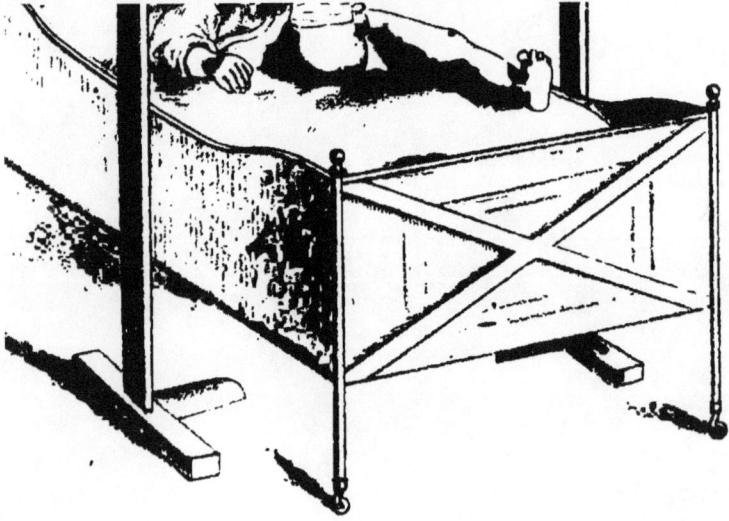

Lith Anst v F Reichhold, Munchen

Explanation of Plate 54.

VERTICAL EXTENSION IN FRACTURES OF THE FEMUR IN CHILDREN.

While the *adhesive-plaster traction dressing in the treatment of fractures of the femur* is generally well established, the same cannot be said of its minor modifications. Yet it is impossible to heal every fracture of the femur with the simple adhesive-plaster traction dressing in the manner desired, even when rather heavy weights are employed. In some cases all that is necessary to avoid serious displacement is to place the entire limb in moderate flexion and abduction so as to bring the lower fragment into equal position with the upper which is influenced by muscular traction. In other cases lateral traction is required in addition, as a rule traction inward and backward in order to prevent an angle salient forward and outward. This is best effected by a loop of adhesive plaster (sometimes two are needed, acting in different directions), which, being fastened to the prominent portion, produces the desired effect by means of a cord and weight attached to it.

In *children* under five years, sometimes also in those a little older, the method shown in Plate 54, *vertical extension*, is the best. It must not be thought that this can be done only with the auxiliaries at hand in a clinic. I have carried out this method of extension in many cases occurring in the district practice of my Munich policlinic. Three boards, if need be nailed to the side of the bed, suffice for the extension frame; the rest is self-evident. The fear that the fragments will unite with deformity, owing to the great restlessness of the children when placed in this position, is unfounded. When the weight balances that of the leg and exerts moderate traction in addition, good union is to be expected.

Explanation of Plate 55.

FRACTURE OF THE PATELLA.

The causes of unfavorable union of fractures of the patella are many. A prominent part no doubt is taken by the traction of the quadriceps which may raise the upper fragment, while the lower one is fastened to the tuberosity of the tibia by the ligamentum patellæ. Although this is an established fact, it should not be accepted without question as an explanation of the displacement of the fragments or for determining the mode of treatment of these fractures. The traction of the quadriceps is of importance only when *the strong aponeurotic layers passing on both sides close to the patella are likewise divided: it has no effect when the patella alone is severed.*

The anatomist W. Braune has called attention to these relations in his beautiful atlas, and has pointed out the slight displacement in fractures of direct origin (stellar fractures) compared with the wide gaping in transverse fractures of the patella due to muscular traction.

FIGS. 1 and 2 show the same preparation in identical position. In Fig. 1 the patella alone is chiselled through; in Fig. 2 the ligamentous tense tissue adjoining the patella on both sides is likewise severed. In the latter case, owing to the position and a certain traction of the quadriceps, the fragments are markedly displaced (*dislocatio ad longitudinem cum distractione*); in the former case the gaping of the fragments is barely perceptible.

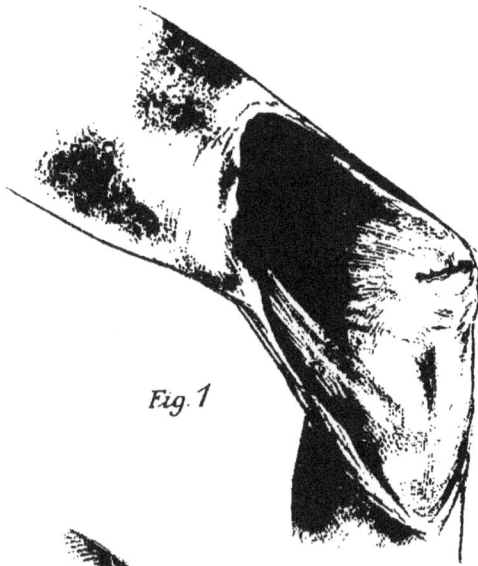

Fig. 1

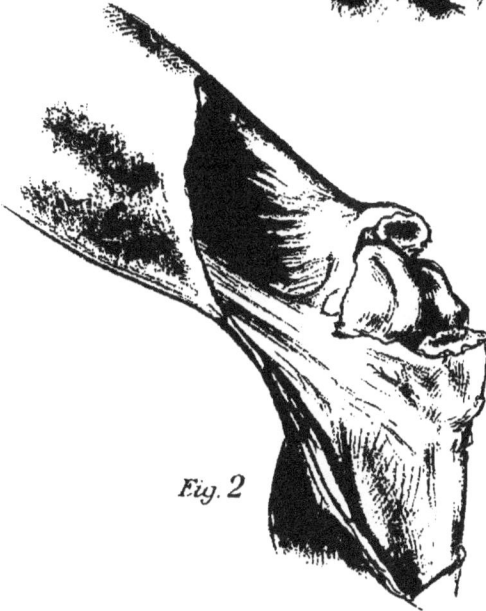

Fig. 2

Lith. Anst. v. F. Reichhold, München.

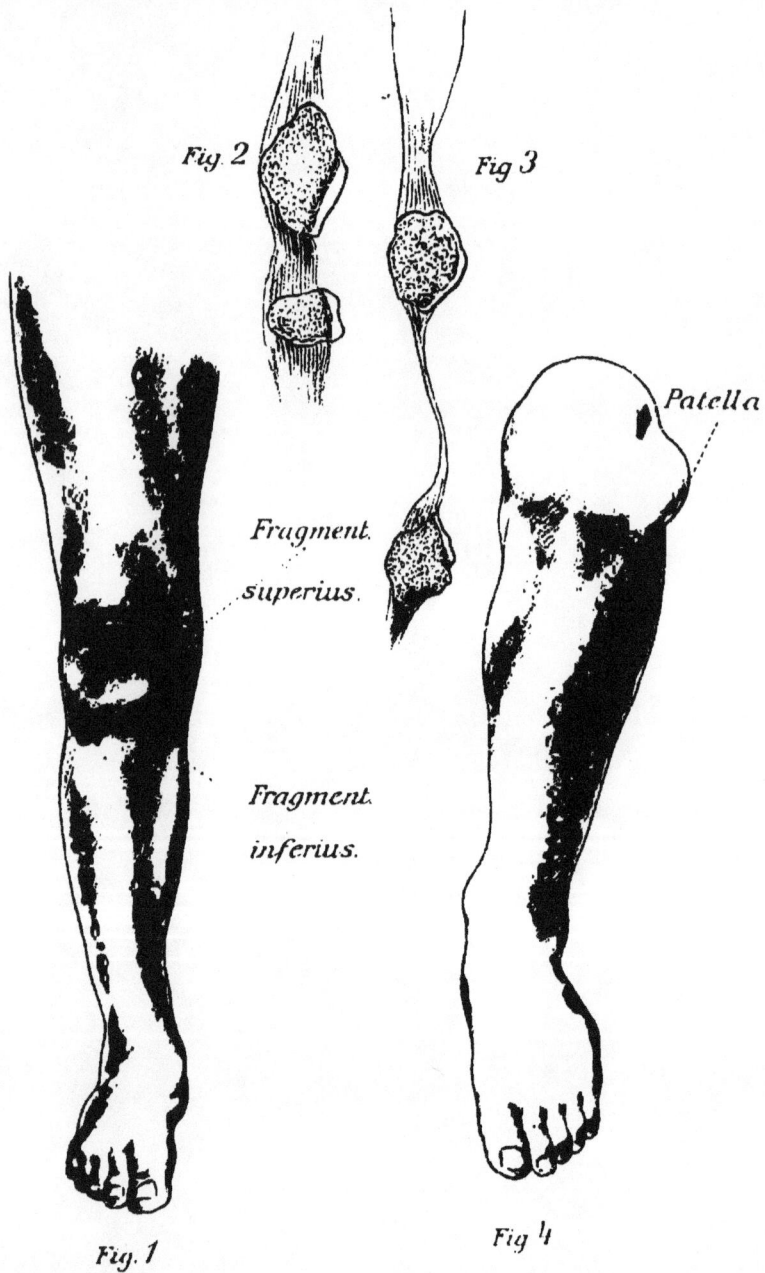

Fig. 2

Fig 3

Patella

Fragment.

superius.

Fragment.

inferius.

Fig. 1

Fig 4

Lith Anst v F Reichhold, Munchen

Explanation of Plate 56.

FIG. 1.—*Old fracture of the patella, united by a broad ligamentous mass* (personal observation). We recognize the two fragments of the patella and the transverse depression between them, which would readily admit two or three fingers.

FIGS. 2 and 3.—Drawings from sketches made by myself at the College of Surgeons, London. The specimens bear the numbers 536 B and 536 F. In these sections we observe the fragments, united in Fig. 2 by a short and broad, in Fig. 3 by a very long and thin, ligamentous mass. We note, too, the cartilaginous investment of the fragments. At the clinic I have sometimes encountered some hesitation as to the condition of the knee-joint in fractures of the patella; I think these illustrations and those on Plate 55 will make it clear to every one that the knee-joint is always implicated in fracture of the patella, that the effusion of blood is within the joint and often enough must be evacuated by puncture with a thick trocar so as to permit an approximation of the fragments. It is clear from the illustrations on Plate 55 that in fracture of the patella the dressing must always be applied with the knee-joint straightened, *i.e.*, with the quadriceps relaxed as much as possible.

FIG. 4.—*Outward luxation of the left patella* in a man aged 29 (personal observation, 1880). This luxation of the patella is the most frequent; of course a slight form of it may be often observed in genu valgum.

Explanation of Plate 57.

FIG. 1.—*Normal course of the epiphyseal lines* at the lower end of the femur and the upper end of the tibia and fibula. Separation of these epiphyses is not very rare both on the femur and on the tibia. But the so-called inflammatory separations of the epiphyses in the course of acute purulent osteomyelitis may be observed more frequently than those of traumatic origin. (Author's collection.)

FIG. 2.—*United stellar fracture of the patella.* Unquestionably this was due to direct injury, and yet the fragments were not markedly displaced. Compare the explanation of Plate 55.

FIG. 3 *a* and *b.*—*Fracture by compression of the tibia* at its upper end. We recognize the course of the lines of fracture at the upper articular surface and the difference in the level of the two halves. If the young woman had survived, a slight genu valgum would probably have remained as a consequence of this injury. The latter was caused by a fall from a loaded hay-wagon, the woman undoubtedly landing on her feet in such a way as to cause a pressure of the condyles of the femur against the upper articular surface of the tibia. The patient died of acute sepsis starting from a fracture by torsion of the same tibia in its lower half, and could not be saved even by exarticulation of the leg. (For a more minute description of this interesting case, see Langenbeck's *Archiv*, Bd. xli., S. 357.) This variety of fracture by compression is quite typical; it causes loosening (possibly lateral to-and-fro motion) at the knee-joint, and is best treated by extension with weights.

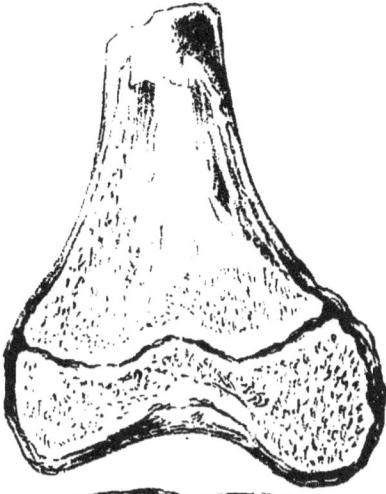

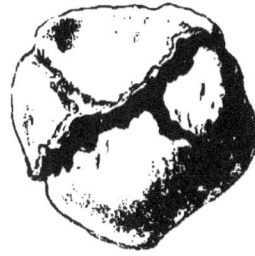

Fig. 2

Fig. 1

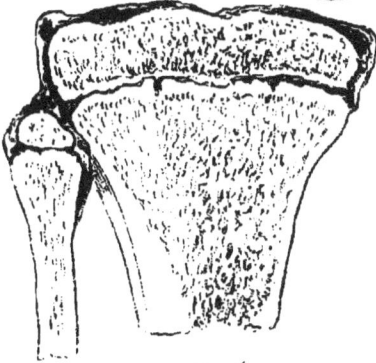

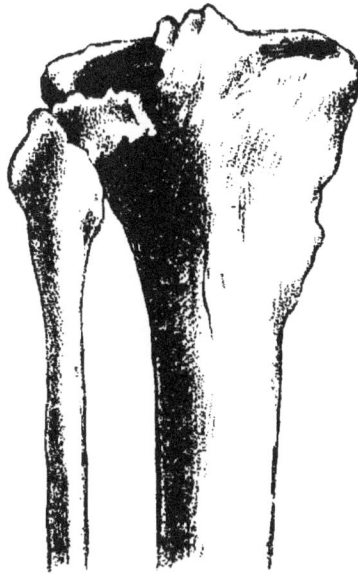

Fig. 3ᵇ

Fig 3ᵃ

Lith. Anst. v. F. Reichhold, München.

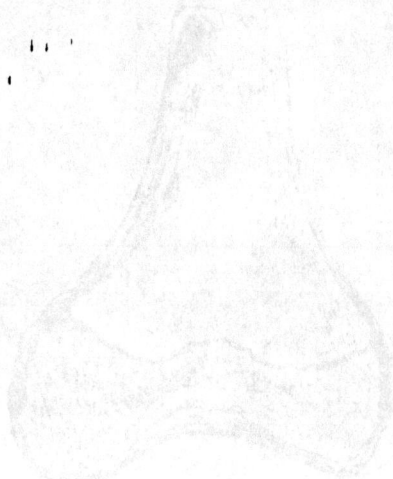

Fig 1

Fig 2

Lith Anst v F Reichhold, München.

Explanation of Plate 58.

FRACTURES OF THE LEG BONES UNITED WITH DEFORMITY.

More attention is paid nowadays to the treatment of fractures than was formerly the case. This is due to the fact that physicians see more frequently than before the final results of fractures and other injuries and give an opinion as to the capacity for work of the former patient. It is astonishing how often long-continued disturbances, even permanent restriction of working-power, are observed after fractures. The physician's art in cases of fracture is not limited to effecting the patient's recovery, but he must bring about complete restoration of the function of the muscles and joints implicated, by appropriate measures such as massage, passive movements, the use of medico-mechanical apparatus, etc.

Figs. 1 and 2 represent deformed legs which were recently presented to us for examination and criticism, as united fractures of the leg bones. They are, I should say, typical deformities after fractures which must be absolutely avoided. The deformity with posterior deviation (Fig. 1) easily results during the application of a plaster-of-Paris dressing when traction in the longitudinal direction of the leg is not powerful, and this outward bending is not prevented by special lifting with slings or the hand while the plaster is hardening.

Explanation of Plate 59.

FRACTURE OF THE TIBIA WITH LUXATION OF THE CAPITULUM OF THE FIBULA.

FIG. 1 was accurately drawn from nature. The patient was a man (Pommering) aged 59, whose tibia was fractured by being run over in January, 1894, and he was admitted to the clinic with a non-consolidated fracture (pseudarthrosis) on May 29th. The injured *tibia was at least two centimetres shorter* than the healthy one, as determined by repeated and very careful mensuration. Accordingly it had to be assumed that the fibula of the injured side must have been either likewise fractured and shortened or luxated. The latter was found to be the case. The fibulæ of both sides were of equal length, but the capitulum of that of the injured side was measurably and visibly dislocated upward, luxated. This could not be completely remedied even under anæsthesia and after the fracture of the tibia had been rendered quite movable; hence the fragments were nailed together and are now consolidated in improved position.

FIG. 2 is exactly the same preparation as that observed in Fig. 1 in the living subject: it is a *fracture of the tibia in the upper half, with considerable displacement of the fragments so as to produce shortening, and with upward luxation of the capitulum of the fibula.*
In the preparation, too, the form of the upper fragment of the tibia is characteristic; it has been compared with the mouthpiece of a clarionet or a duck-bill. This fragment might become very troublesome by its prominence and its pressure against the thin integument (gangrene). Many complicated measures have been recommended (for instance, Malgaigne's thorn); the best is a very careful reposition (possibly under anæsthesia) and suitable dressing in extended position. (Personal observation.)

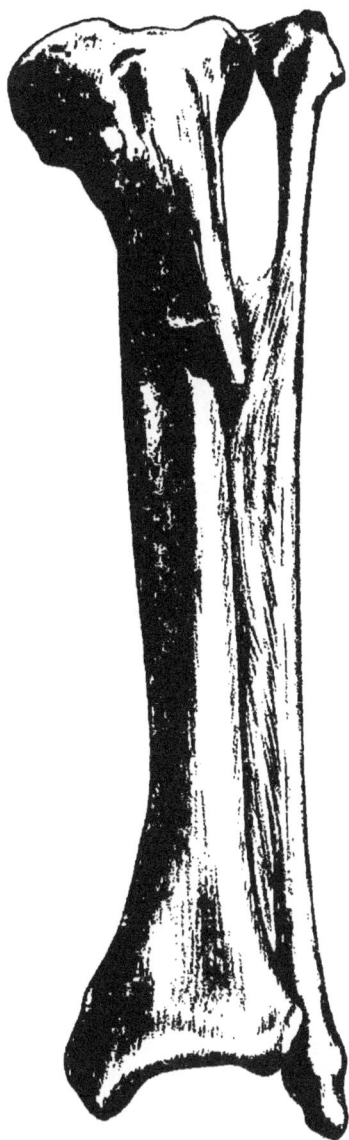

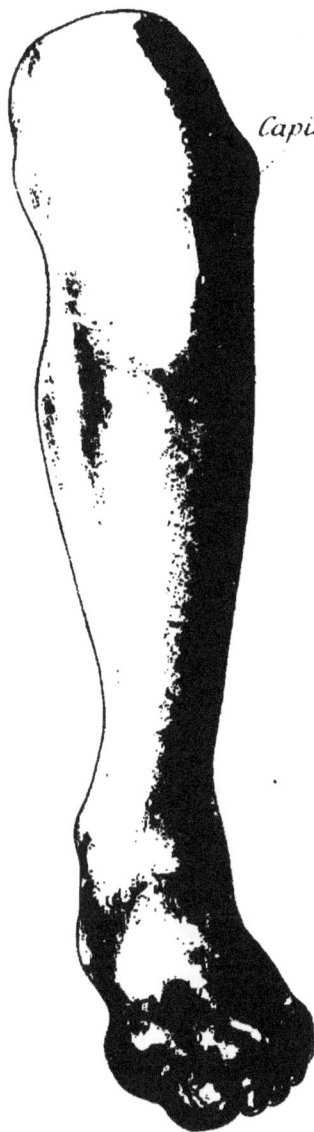

Capit fibul

Fig. 2

Fig. 1

11

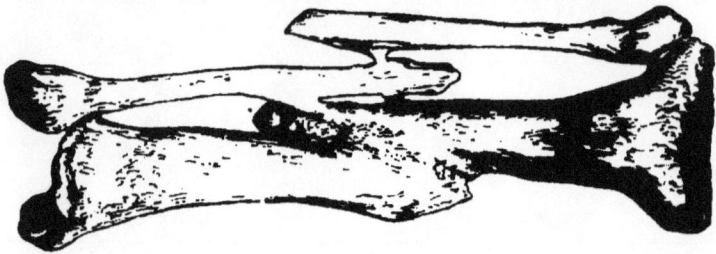

Fig. 1

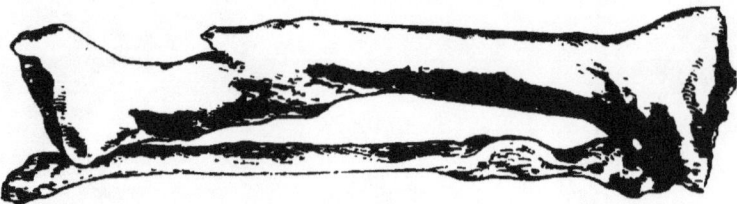

Fig 2

Fig 3

Fig 4

Lith Anst v F Reichhold München

Explanation of Plate 60.

FIG. 1.—*Fracture of the leg bones united with marked displacement of the fragments.* The tibia and fibula are broken at about the same level, have been similarly displaced, and have united by thick callus formation which moreover joins tibia and fibula together. (Pathological Institute in Berlin; after Wolff, "Transformation der Knochen," Plate VII., Fig. 48.)

FIG. 2.—*Fracture of the leg bones united with slight displacement of the fragments.* The tibia is fractured in its lower, the fibula in its upper half.

It is quite correct that one of the bones, if it remains intact, forms a kind of splint for the other which is broken. It seems that the location of the fracture at different levels of the two bones likewise exerts a favorable influence by preventing marked displacement. The difference between the displacements in Figs. 1 and 2 may in part be due to this fact, *i.e.*, to a certain fixation effected by the interosseus membrane. (Author's collection.)

FIG. 3.—*Supramalleolar fracture of both leg bones united with marked displacement.* The displacement is in the sense of a severe pes valgus. (Author's collection.)

FIG. 4.—*Recent fracture by torsion at the lower end of the tibia.* (Author's collection.)

Explanation of Plate 61.

Typical Malleolar Fracture.

This specimen is an artificially produced malleolar fracture; the leg was sawed through in the frontal plane as shown, and the preparation thus obtained was then drawn from behind. On the foot we recognize at once the great toe which serves as a landmark. The artificial production of this fracture in the cadaver is not difficult; it succeeds almost invariably when the leg rests on its outer side, the region immediately above the external malleolus corresponding to the edge of the table; then a sudden strong pressure upon the foot suffices to effect *separation of the internal malleolus and fracture of the fibula above the external malleolus* in the typical form shown in the illustration. Every physician should gain a correct conception, in this manner, of this important injury; it will be more valuable to him than all descriptions.

In the illustration we see the tibia in frontal longitudinal section; the fibula was not touched by the saw, but the posterior portion of the astragalus and the calcaneus were severed. The internal malleolus is separated and displaced downward and outward. The most important change is shown by the fibula at the point of its fracture—an angle open outward. This is the essential cause of the occurrence of traumatic pes valgus after such fractures. This infraction of the fibula, the outward displacement of the astragalus and of the entire foot, the sinking of the inner margin of the foot—all this can be well seen.

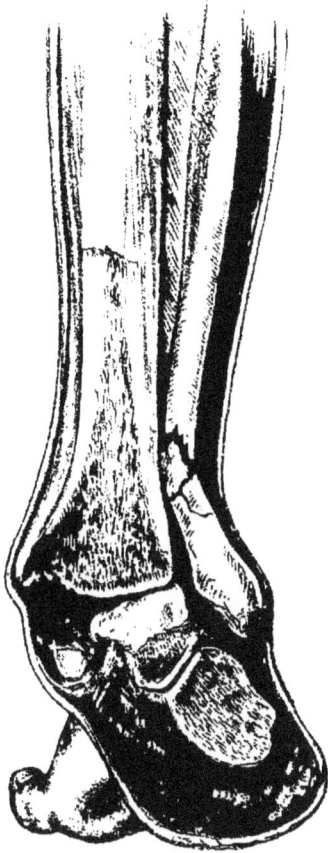

Lith Anst v F Reichhold, Munchen

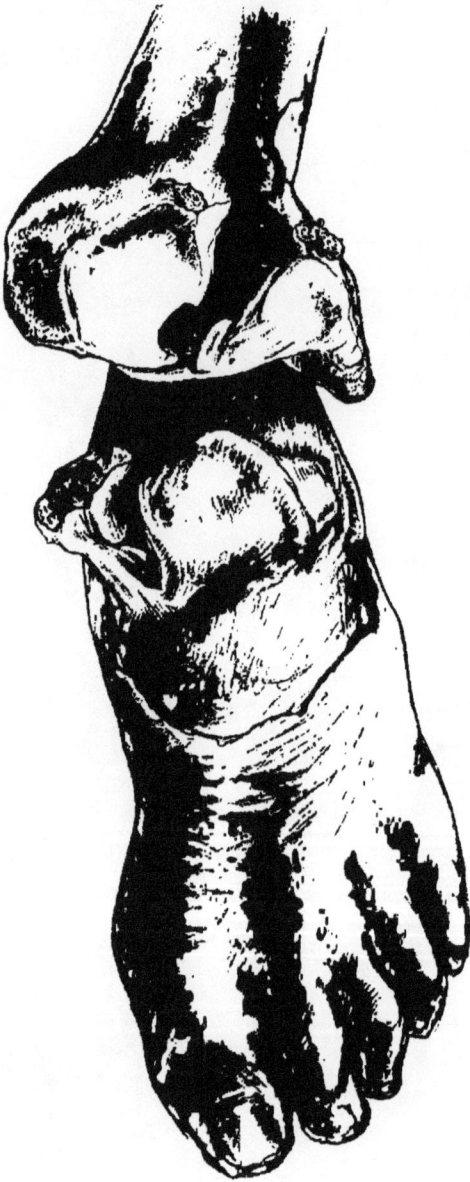

Lith Anst / F Reichhold, München

Explanation of Plate 62.

TYPICAL MALLEOLAR FRACTURE.

There is no injury in which the character of a fracture by traction is more clearly marked than in this. The fact is well established that the ligaments are often firmer, and better able to resist a sudden traction, than the bone. Under such circumstances portions of the bones are torn off by the respective ligaments. The present plate shows the internal malleolus torn off by the strong deltoid ligament during violent abduction of the foot. At the junction of the tibia and fibula we also observe that small particles are torn away from the tibia in front and behind; the anterior particle by the anterior tibio-fibular ligament, the posterior by the posterior tibio-fibular ligament. The traction of the latter ligaments, however, was rendered possible only after the malleolus was fractured by the severe infraction of the external malleolus; this fracture again was effected by the pressure of the foot, especially the astragalus, against the external malleolus. The separation of the internal malleolus, therefore, is the primary lesion.

In some cases only the internal malleolus is separated, in others the severest grade of malleolar fracture results as illustrated in the plate. The separation of particles of bone at the fibular side of the tibia is always observed, even when the fibula is remarkably infracted and displaced.

This preparation likewise was produced artificially. The view into the ankle-joint injured by the malleolar fracture afforded by the drawing is particularly instructive.

Explanation of Plate 63.

FRACTURE OF THE ANKLE WITH DISPLACEMENT OF THE FRAGMENTS.

FIG. 1.—*Traumatic pes valgus*, caused by fracture of the ankle. We observe the infraction of the fibula above the external malleolus, the widening of the ankle region, the descent of the inner margin of the foot (flat-foot position). The illustration represents the condition of a man admitted with this deformity after fracture of the ankle, who was treated by osteotomy at the seat of the fracture of the fibula and forcible reposition; the function of the ankle-joint became very good by the use of massage and all possible mechanical auxiliaries.

FIG. 2.—*Bilateral, typical fracture of the ankle with backward subluxation of the foot.* The drawing was made from the photograph of a man aged 52, who had fallen a distance of two metres and suffered this injury. Reposition succeeded without difficulty; perfect recovery.

FIG. 3.—*Epiphyseal lines of the tibia and fibula* at their lower end. Traumatic separation of the epiphyses may occur also at this point, though such injuries are rare. The treatment, and especially the reposition, will have to be carried out on the general principles governing fractures of the ankle.

72

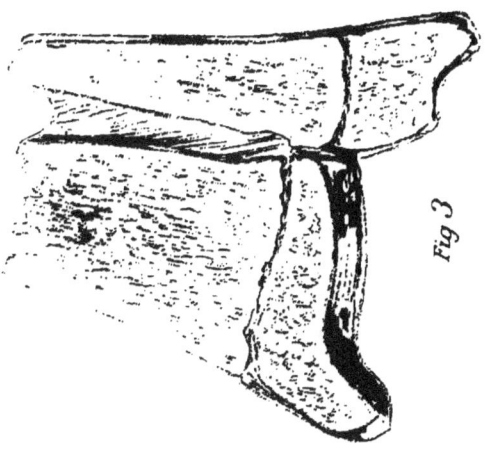

Fig 3

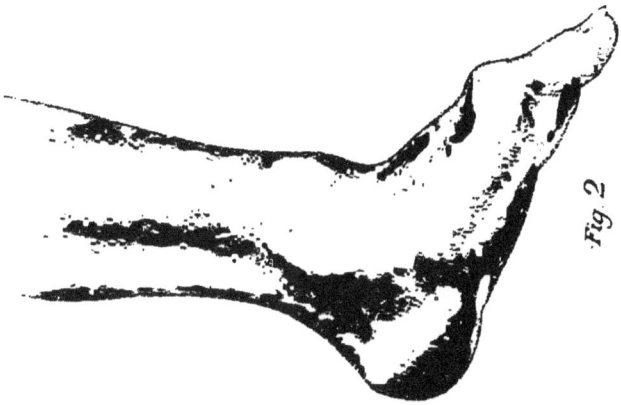

Fig 2

Fig.1

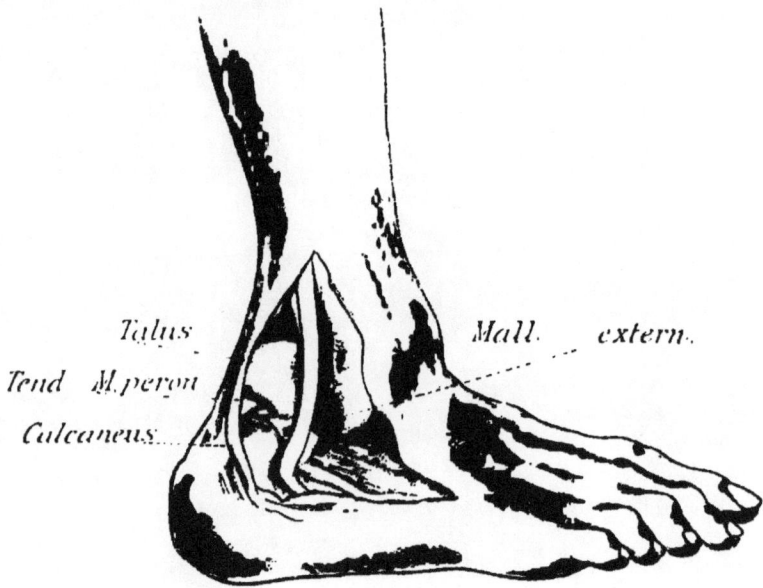

Talus

Tend. M.peron

Calcaneus

Mall. extern.

Fig. 1

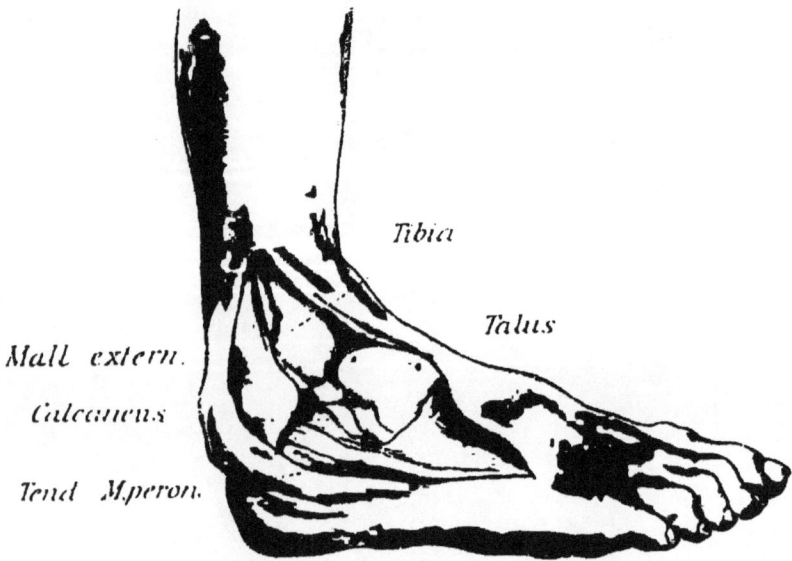

Tibia

Talus

Mall. extern.

Calcaneus

Tend. M.peron.

Fig 2

Lith Anst. v F Reichhold. Munchen

Explanation of Plate 64.

BACKWARD AND FORWARD LUXATION OF THE FOOT.

The two preparation here shown were produced artificially. Pure luxations at the ankle-joint (astragalo-crural articulation) are infrequent injuries; this is true also of luxation in the astragalo-tarsal joint (subtalic luxation) and of isolated luxation of the astragalus. Combinations of fractures with luxations are here relatively more frequent.

FIG. 1.—*Backward luxation of the foot.* We observe the astragalus behind the external malleolus and the interposed peroneal tendons. The foot is shortened in a characteristic way: a deep transverse furrow passes from one malleolus to the other; the heel portion of the foot appears markedly elongated.

FIG. 2.—*Forward luxation of the foot.* The astragalus is situated in front of the leg bones, both of which are visible in the illustration. The heel is conspicuously shortened, the entire foot is enormously elongated, and the skin on the dorsum of the foot is stretched.

The reduction of these injuries is usually not difficult under anæsthesia. Should obstruction be encountered operative reduction should be performed, as has recently been done with success by v. Bergmann in a case of isolated luxation of the astragalus.

Explanation of Plate 54

BACKWARD AND FORWARD DISLOCATION OF FOOT

V. Fractures and Luxations of the Lower Extremity.

1. PELVIS (PLATE 45).

SOLUTION of continuity occurs at the pelvis only in consequence of powerful forces, such as a fall from a great height, the impact of large and heavy objects, the caving in of an excavation, etc. In this way fractures and diastases of the symphyses between the pelvic bones may result. The latter are even rarer than fractures. Their occurrence at the pubic and sacro-iliac symphyses necessitates the laceration of the exceedingly firm ligamentous connections; the diastasis of the Y-shaped cartilage in the acetabulum (compare Plate 45, Fig. 2) is possible only as an incidental injury with other separations of the bones constituting the pelvic ring. It is only when such a diastasis is complicated with pronounced displacement of the parts on each other that the diagnosis can be made with certainty. In other cases, especially those in which the sacro-iliac symphysis is implicated, merely the signs of a severe distorsion are present, but they suffice for making us appreciate the importance of the lesion when the cause of the injury is known. The treatment is based on general principles.

Pelvic fractures should be differentiated clinically according to whether isolated parts of the pelvis are fractured or the continuity of the pelvic ring is

actually broken.   In the former case there may be a
fracture of a portion of the venter of the ilium, a fracture of the sacrum or coccyx, or of the tuberosity of
the ischium.   The separated parts sometimes appear
abnormally movable on direct examination, together
with crepitation and displacement.   Incidental injuries are rarely present with these separations.   The
treatment aims at union in the most correct position
possible, though moderate changes of form produce
no ill effect.

Fractures of the pelvic ring are much more important.   On the one hand severe traumata must have
acted in breaking the continuity of the pelvic ring,
and on the other hand, and partly in consequence of
this fact, incidental injuries are not rare.   Among
the latter are lesions of the sciatic and other nerves,
of the femoral vessels, of the bladder and of the rectum, which are seldom observed; but relatively frequent, and of the greatest practical importance, are
lesions of the urethra in pelvic fractures in men; they
manifest themselves by the escape of blood from the
urethra and the admixture of blood in the urine.
The introduction of a catheter is of importance not
only for diagnostic purposes but also therapeutically
(catheter à demeure).   Should catheterization fail,
the danger will be imminent of the development of
an infiltration of urine in the surrounding cellular
tissue, with all its evil results of a fatal gangrene and
sepsis.   Such cases, therefore, require an immediate
free incision from without in the way of an external
urethrotomy extending to the cellular tissue, which
is generally greatly infiltrated with blood, about the
urethra in the region of the bulb, and especially the

membranous part. The performance of true urethrotomy is often very difficult and sometimes impossible, so that the high operation with the so-called retrograde catheterization would be indicated. Without the auxiliaries of a hospital this operation can hardly be completed; but the physician may be expected to make an incision extending into the cellular tissue about the urethra, and at least to diagnose the severe injury, so that appropriate treatment may be instituted.

The forms of fracture of the pelvic ring are very manifold. Aside from the influence of the spinal column and the thighs upon the pelvis, we meet mainly with compressions of the pelvis from before backward (for instance, a wagon-wheel passing over a person lying on the back, the fall of a horse upon its rider, etc.), or in a lateral direction. These relations have also been tested by experiments. On compression from before backward the anterior pelvic wall first breaks down (lines of fracture through the upper and lower surroundings of the obturator foramen on both sides) and then follows a separation of the sacro-iliac symphysis or a fracture alongside of it in the sacrum. On lateral compression a fracture likewise occurs first at the anterior portion which possesses the least resistance (about the symphysis pubis) through the obturator foramen, then also a fracture through the ilium beside the sacro-iliac symphysis, unless the ligamentous apparatus of the latter yields. Thus one half of the pelvis may be fractured in front and behind simultaneously—the so-called double vertical fracture of Malgaigne. Numerous other lines of fracture may arise when the pelvis is compressed diagonally. In the living subject the causes of pelvic fractures as a rule are so powerful and manifold that the pelvis does not fracture in such a typical manner, but does so at many points; hence we find specimens with fifteen or twenty or even more separate lines of fracture and fissures.

During the examination it is useful to attempt to compress the pelvis by means of the hands applied upon the crests of the ilia. When fracture is present this causes a violent pain at its seat, sometimes also abnormal mobility and crepitation.

The prognosis depends upon the concomitant injuries; in their absence union may be expected.

Treatment.—Suitable position (water bed, pillows of millet chaff), sometimes on a kind of portable frame as in fracture of the vertebræ, so as to obviate movements of the patient for the purpose of defecation. A belt-shaped dressing around the pelvis is often useful and especially grateful to the patient. In fracture involving the acetabulum, careful mobilization of the hip-joint.

2. HIP-JOINT.

Luxations in the hip-joint are among the rarer injuries; a powerful force is required to produce them. The most important varieties are backward and forward luxations; others are much less common. Since the investigations of Bigelow, of Boston, the determining factor in the mechanism and the fixation of the luxated bone is the ileo-femoral or Bertini's ligament which is preserved in all regular luxations; only when it is torn is an irregular dislocation without characteristic symptoms possible.

A. Backward Luxation. Luxatio Postica s. Retrocotyloidea (Plates 46, 47).

When in the cadaver the thigh in flexed and slightly adducted position is rotated inward, the joint capsule on its posterior surface is tensely stretched; when the motion is continued the neck of the femur presses against the anterior margin of the acetabulum which forms the fulcrum that permits of the exertion of an enormous force by means of the long lever (shaft of the femur) upon the short lever (head of the femur). The head is crowded against the capsule, the latter tears on its posterior surface, the head leaves its articular connection (the ligamentum teres being lacerated) and the posterior luxation is effected.

In the living subject the backward dislocation comes about in this manner; whether due to a movement of the thigh (more rarely) or of the trunk or pelvis when the thigh is fixed (more frequently).

We distinguish an iliac and a sciatic luxation. In the former the head rests upon the ilium; in the latter it is lower, on the upper portion of the ischium. An important anatomical difference consists in the position of the tendon of the obturator internus muscle with reference to the head of the femur: in iliac luxation the head of the femur is above, in sciatic luxation below this tendon.

Symptoms.—In backward luxations the thigh is rotated inward and elastically fixed in more or less marked flexion and adduction. When the patient is in dorsal decubitus we recognize this position and a shortening of the limb, which is greater in iliac luxation and less in sciatic luxation. The shortening

may be determined by mensuration, starting from the
anterior superior spine and passing to a point at the
knee-joint (say the lower edge of the patella or the line
of the knee), both thighs being in symmetrical position with reference to the pelvis.   Roughly, the shortening will be very distinct when the two thighs are
placed in right-angled flexion symmetrically to the
pelvis and compared with each other; the pelvis must
be quite horizontal and both anterior superior spines
at the same level.   In backward luxation the knees
are not at an equal height; that of the injured side
being considerably lower because the corresponding
femur is dislocated backward on the pelvis.   This
procedure is especially appropriate for examination
under anæsthesia.

The dislocation can also be accurately measured in
the coxal region.   Under normal conditions the line
passing from the anterior superior spine to the tuberosity of the ischium across the gluteal region (for instance, by a tape), the thigh being flexed, strikes exactly the tip of the great trochanter.   This is called
the Roser-Nelaton line.   In backward luxation the
upper end of the femur is dislocated upward and therefore the trochanter is elevated above this line; it is
found on making this examination, the patient lying
on the healthy side, that it is at a greater or lesser
height above the normal, and thus we can deduce the
position of the head of the femur, provided its connection with the neck and shaft of the bone is intact.

During this examination the inward rotation becomes manifest in so far as under normal conditions,
in a position midway between outward and inward
rotation, the tip of the trochanter is about in the cen-

tre of the Roser-Nelaton line. The position of the trochanter forward of the centre of this line indicates the inward rotation of the thigh which is never absent in the regular backward luxations,* and thus points to the location of the head of the femur behind the acetabulum.

This examination, too, will hardly be feasible as a rule without anæsthesia. The amount of the displacement may also be estimated in a more simple way if the patient is placed on his back and the physician puts his thumbs, in the most symmetrical position possible, on the two anterior superior spines and thence determines with the index finger the location of the tip of the trochanter on each side; in this way he can sometimes measure approximately the distance of the two bony points by the number of fingers which can be inserted between them, and thus roughly estimate the position of the tip of the trochanter with reference to the pelvis.

The presence of the head of the femur in its abnormal position is not always demonstrable under the thick gluteal muscles, particularly when the swelling is great and anæsthesia is not induced.

Active movements are completely arrested. Passively we can effect a slight increase of the perverse position in the way of adduction and inward rotation, but only by inflicting great pain; on attempting abduction and outward rotation of the thigh we

---

* There is a backward luxation with outward rotation of the thigh; this is very rare and can occur only when there is a laceration of at least the outer crus of Bertini's ligament and the joint capsule is extensively torn.

find the characteristic elastic resistance which is due mainly to the tension of Bertini's ligament.

Treatment.—It has been repeatedly stated that anæsthesia cannot well be dispensed with in the examination; of course when the diagnosis has been made reduction immediately follows. To this end it is desirable in all cases to anæsthetize the patient profoundly and to place him on the floor (upon a blanket or mattress). Then the affected limb is raised until the thigh is vertical and the manipulations are made with the leg bent at a right angle in the knee-joint. Sometimes a simple upward traction suffices to effect reduction, of course only when the head is close to the posterior margin of the acetabulum. If the head is dislocated farther it may on simple traction be jammed against the margin of the acetabulum, and it is readily understood that in this way the obstruction increases when the thigh is at the same time in abduction which *a priori* appears very useful for the reduction. Hence it will be clear why it is advised that traction should be made with the thigh in adduction, because the head glides more easily over the margin of the acetabulum —that is to say, traction in the adducted position with some inward rotation. If this fails traction should also be attempted in abduction position with outward rotation; during this manipulation, by the way, the head of the femur may also deviate in such a manner that it passes around the anterior surface of the articulation (so-called circumduction). Since such secondary movements of the head of the femur, therefore, are not excluded, we cannot always deduce the form of laceration of the capsule from the position of the head. The capsule, which may be lacerated longitudinally or transversely, sometimes forms a true obstruction to reduction, which can only be overcome by operation (incision). By this means I have reduced a backward luxation of several weeks' standing in a child, with preservation of complete mobility. In very old cases resection of the hip may be performed, or relinquishing mobilization of the luxated head, a subtrochanteric osteotomy with a view to improve the perverse position.

B. FORWARD LUXATION. LUXATIO ANTICA s. PRÆCOTYLOIDEA (PLATES 46, 48):

Forward luxations are rarer than backward; referring to the latter, I may be more concise in my remarks.

The artificial production of a forward luxation succeeds by outward rotation and abduction. The capsule tears on its anterior side; more superiorly with a suprapubic luxation when the thigh is extended at the same time (hyperextension); more inferiorly with an infrapubic luxation when the thigh is flexed.

In the living subject forward luxation occurs in the same way or by a corresponding displacement of the pelvis when the thigh is fixed.

In all forward luxations the lower extremity is in marked outward rotation* and abduction. The degree of flexion varies: in suprapubic luxation it is slight, sometimes even extension is present; in infrapubic luxation flexion is never absent, and is more

---

* Only when the head of the femur was dislocated upward into the pelvis has an inward rotation been observed; this is extremely rare.

pronounced in proportion as the head of the femur is dislocated farther inward (a result of the tension of Bertini's ligament).

In suprapubic luxation the head is directly palpable in the inguinal region; it may be either close to the margin of the acetabulum (*luxatio ileo-pectinea* with very slight abduction) or upon the os pubis (*luxatio pubica*). The femoral artery is sometimes lifted up by the head of the femur; pains are present in the region of the crural nerve. At times the patient can still bear the weight of his body on the injured leg.

In the case of an infrapubic luxation there will be, besides outward rotation, a more pronounced abduction and flexion. We distinguished two forms: obturator luxation when the head is in the region of the obturator foramen, and perineal luxation, which is very rare, when the head is dislocated as far as the ascending ramus of the ischium. In obturator luxation the head is hidden in the depth and not easily felt, the prominence of the trochanter is absent, and the leg is elastically fixed in its abnormal position.

In the diagnosis fracture of the neck of the femur is excluded by the fact that though the thigh is likewise shortened and rotated outward in this injury, the elastic fixation characteristic of the luxation is absent; the thigh may be straightened without difficulty, though it drops back into outward rotation; and other movements are not prevented as in luxation.

In reducing the suprapubic luxation traction in hyperextension may first be necessary so as to approximate the head to the acetabulum; during this step the patient must be suitably placed on a table. Otherwise the same rule applies as in backward luxation, that the patient must be anæsthetized, placed on the floor, and the manipulations made with the leg more or less flexed. Inward rotation followed by adduction generally succeeds. Circumduction of the head about the margin of the acetabulum (see above) may be avoided by simultaneous traction on the thigh.

C. RARE LUXATIONS AT THE HIP-JOINT.

Downward luxation (infracotyloid) is very rare; the head of the femur is at the lower margin of the acetabulum, the thigh is elongated; marked flexion is never absent and slight abduction is usually present; rotation is immaterial. The injury may result from forced abduction. Reduction by traction on the flexed thigh.

Upward luxation (supracotyloid) is likewise very rare. The head is at the anterior inferior spine and may be felt directly as a spherical prominence. The thigh is extended, slightly rotated outward and adducted, and considerably shortened. Reduction by flexion and inward rotation.

The name central is applied to the extremely rare luxation of the head of the femur into the pelvis through the comminuted acetabulum. This observation is of interest on account of its analogy with the skull (fracture of the base of the skull by the lower maxilla).

3. THIGH.

A. FRACTURES AT THE UPPER END OF THE FEMUR (PLATES 49, 50).

a. Fracture of the neck of the femur is a typical injury which is of great practical importance. We dis-

tinguish the so-called intracapsular and extracapsular
fractures of the neck of the femur, according to the
direction of the line of fracture, whether nearer to the
junction of the head and neck of the femur or to the
trochanter.   As the neck of the femur is situated
largely within the joint capsule, the occurrence of a
pure intracapsular fracture is indeed possible; extracapsular fractures, however, are usually in part at
least within the capsule, and therefore are "mixed."

Fractures of the neck of the femur result as a rule
from a fall upon the hip, *i.e.*, upon the trochanter
(these are chiefly extracapsular and complicated with
impaction of the neck into the trochanter) ; they may
also be due to a fall upon the extended leg or knee
(often intracapsular), and even to tension and traction
of Bertini's ligament during extensive rotary movements (fracture by traction).   It is in the last-named
manner, especially in the cadavers of old subjects, that
fractures of the neck of the femur often result when
the attempt is made to produce an artificial luxation
of the hip-joint.

The frequent occurrence of these fractures in old
people is due to the fragility of the bones, which is
often especially pronounced at the upper end of the
femur.   Under normal conditions, as is well known,
this part is very firm and quite equal to the task of
bearing the weight of the body.   We are familiar
with the principles of the architecture of the osseous
trabeculæ, which answers every mathematical or
mechanical requirement and combines the highest
stability with the least amount of bone substance.
With advancing age the osseous trabeculæ become
scantier, the intermediate cavities filled with fat grow

larger, the bone itself loses some organic ingredients; in this way an osteoporosis results, which, by the way, occurs generally·earlier in women than in men. This fact explains the more frequent occurrence of fractures of the neck of the femur in women.

If the pressure which produces the fracture acts approximately in the longitudinal direction of the neck of the femur, there is apt to be an impaction (gomphosis) of the fragments. In that event the thinner and firmer portion of the neck is wedged into the cancellated structure of the head (in intracapsular fractures) or into that of the trochanter (in extracapsular fractures). The impaction is of practical importance because its symptomatology is altered, and because the rule applies that it is not to be disturbed. Clinically, impacted fractures of this kind resemble incomplete fractures (infractions) of the neck of the femur; sometimes they merely present an inflexion of one side of the neck of the femur, which produces a change of its direction and length, a more or less obtuse-angled junction between neck and shaft, and consequently a higher position of the trochanter.

Symptoms.—Fracture of the neck of the femur should always be suspected when a person of advanced age is unable to walk in consequence of a fall and the injured leg is shortened and rotated outward. The differentiation from forward luxation of the femur is made by the fact that the thigh is fixed in outward rotation but not elastically; it can easily be straightened but immediately drops back outward. The outward rotation of the thigh is slighter, often quite inconsiderable, in impacted or incomplete fracture; very pronounced in the ordinary loose fracture.

Otherwise the thigh lies straight, without abduction or adduction and without flexion.

A matter of great importance is the higher position of the trochanter, which is to be demonstrated in the same way as described under posterior luxation in the hip-joint. When the thighs are in symmetrical position to the pelvis the tape shows that the distance from the anterior superior spine to the knee is often considerably shortened. The proof that the tip of the trochanter is situated above the Roser-Nelaton line the same distance that the thigh is shortened indicates that the femur is otherwise intact, and that the cause of the shortening is to be sought at the neck of the femur or in the hip-joint. This may be verified by showing that the distance from the tip of the trochanter to the knee, measured symmetrically, is equal. The shortening is the result of muscular traction acting upon the shaft of the femur (including the trochanter).

Movements of the injured thigh are possible in all directions, though they are painful. These movemens are associated with crepitation when the fragments are not greatly displaced, but are still in contact. In rotating the extended thigh one landmark is sometimes very distinct, whose explanation is clear *a priori*; namely, in extracapsular fractures the shaft of the femur turns about its longitudinal axis; in intracapsular fractures about a radius whose length corresponds to the intact portion of the neck which has retained its connection with the femur.

Impacted fractures are, as may be gathered from what has been said above, distinguished by slighter shortening and less outward rotation of the thigh, by the absence of all crepitation, by less displacement, and finally by the fact that rotary movements at the hip-joint have for a radius the neck of the femur.

Treatment.—As the patients are usually old people, good nutrition and the preservation or improvement of the general health are of great importance. The occurrence of an asthenic hypostatic pneumonia is but too often fatal; hence aside from suitable nourishment frequent change of position (as much as possible), an occasional sitting up, and deep breathing are indicated; early walking about by means of ambulatory splints is especially useful in these cases.

Extracapsular fractures as a rule unite by abundant callus; for bone formation after fractures or osteotomies in the trochanteric region in general is usually very extensive. Intracapsular fractures rarely heal by osseous union because the head is poorly nourished, being normally connected only with the ligamentum teres; sometimes a ligamentous union results, more often a true pseudarthrosis: the head fixed in the acetabulum and the remnant of the neck by its to-and-fro motion abrading each other so that approximately congruent surfaces are in contact.

When the diagnosis of an impacted or incomplete fracture has been correctly made, the limb must be placed at rest and used with caution until the bone has recovered the firmness necessary to its function. Even weeks after the injury the impaction may be loosened and the fragments become displaced; in these cases, therefore, great care is required.

In the ordinary fractures of the neck of the femur the most exact possible reposition of the fragments (extension and inward rotation) is necessary. Then it is best to apply a correct adhesive-plaster extension

dressing with permanent extension by weights and
pulleys according to the rules of surgical technique;
the foot rests on a well-padded sliding board (Volkmann's) so as to counteract at the same time the outward rotation of the leg.   As a rule a weight of twelve
to fifteen pounds suffices to keep the fragments in
good position.   One advantage of this dressing is
that it permits comparatively great mobility to the
patient: a semi-recumbent position in the bed, even
some degree of sitting up, are possible without harm
or pain.   No other splint is required.   That plaster-
of-Paris and splint dressings may likewise be used is
self-evident.   For these cases in particular the new
ambulatory splints (of Thomas, Liermann, and
Bruns) are applicable; here the tuberosity of the
ischium forms the fixed point, and even permanent
extension by rubber straps can be employed, to be
changed at night to traction by weights.

The attempt to fix the fragments by operation (for
instance, the insertion of a gimlet from without) is
indicated only in special cases.

The final result is usually not very brilliant.   As
the patients are old and feeble, we must be satisfied if
they learn to walk again after six or eight weeks, and
subsequently go about with the aid of a cane.

b. Isolated fracture of the great trochanter is a very
rare injury caused by direct force, and is marked by
the easily comprehended displacement of the broken
prominence (*dislocatio ad longitudinem cum distractione*).   The dislocated fragment can be felt behind and above through the glutæi; between it and
the femur is a wide diastasis.   The simplest treatment
would be the nailing of the fragment after the com-

pletest possible reposition, which will be facilitated by abduction of the thigh.

B. FRACTURES OF THE DIAPHYSIS OF THE FEMUR (PLATES 51, 53, 54, 57).

Fractures in the middle portion of the diaphysis are frequent, especially those somewhat above the middle. After referring to the plates enumerated above, a brief description will suffice here. While a portion of the fractures of the diaphysis result from torsion (oblique and longitudinal fractures), the majority are due to flexion by some direct force (run-over accident).

These fractures are also frequent in children, in whom they are not seldom relatively favorable by the fact that the thick periosteum is preserved, whereby a material displacement of the fragments is prevented. In adults the displacement of the fragments is as a rule quite considerable; the line of fracture usually runs obliquely, so that a displacement easily results from the traction of the powerful muscles, which acts in the main in the longitudinal direction. Abnormal mobility is generally readily demonstrated. Crepitation is usually quite distinct, and it must be emphasized that this symptom should be actually demonstrated; for if crepitation is absent the fragments are presumably much displaced or soft parts are interposed; crepitation must be elicited in order to insure contact of the fractured surfaces and correct union. The shortening caused by the longitudinal displacement of the fragments is always readily determined by mensuration from the knee (lower margin of the patella or the line of the knee) to the trochanter or, better, to the anterior superior spine, the thighs being in symmetrical position.

Fractures above the middle are as a rule distinguished by a typical displacement, which unfortunately is found but too often after union, so as to require renewed surgical special interference on account of the angular position. A fracture of the diaphysis of the femur above the middle united with deformity presents at the site of the fracture an angular projection outward and forward. In other words, the upper fragment, under the influence of the muscles inserted at the great trochanter, is in a position of flexion (by the ileo-psoas) and abduction (by the glutæi). The lower fragment is approximated to the upper (over-riding) at the point of fracture, while the lower part of the shaft is still acted upon by the adductors. In this way the angular position is produced.

Treatment.—The treatment of fractures of the diaphysis has become very simple since the introduction of the adhesive-plaster extension dressing with permanent extension by weights. By this means the traction of the muscles is successfully counteracted and deformities are prevented. But it would be an error to suppose this treatment to be devoid of trouble; in the first place the dressing must be applied with absolute correctness, it must cause no pressure anywhere, must adhere by broad surfaces, and be strong enough to bear a weight of from twenty to twenty-five pounds. For this purpose strips of plaster from thick sail-cloth are used. In order to prevent friction of the leg upon the mattress a sliding foot-board (Volkmann's) is employed, which permits at the same time the retention of the foot in a definite position, if necessary in slight inward rotation. Counter-extension is best effected by raising the foot of the bed by blocks of wood or bricks, and by furnishing the healthy foot a firm point of support by a block of wood placed in the bed.

After the patient has been thus bedded, the next task of the surgeon is to keep in view the point of fracture; this is facilitated by the fact that it is open to inspection at any time. But the displacement cannot always be clearly felt under the thick muscles; hence from time to time exact mensuration of the thigh and comparison with the healthy side are required. The measurement of the injured thigh, say from the lower margin of the patella (through the dressing) to the anterior superior spine, is not difficult. The measurement of the healthy limb, however, must be made in precisely symmetrical position. To this end we first determine the horizontal axis of the pelvis, the line connecting the two anterior superior spines; a line extending vertically downward from its centre (for instance, a cord or tape) permits us to estimate the degree of abduction of the injured thigh, which of course remains undisturbed in the dressing, and to have the healthy limb placed in equal abduction and flexion by an assistant. After that mensuration may be effected between the corresponding terminal points and the results of the two sides compared.

While this description sounds complicated, its performance is simple to the experienced, and is important for obtaining favorable results.

Not rarely this examination proves that simple extension, even with heavy weights, is insufficient. In such severe cases the old rule of bringing the lower fragment into the same position as that occupied by the upper is to be observed : the injured thigh is placed in moderate abduction and flexion while extension by weights is applied. In children vertical extension (Plate 54) is an excellent method. In newborn children and infants the simplest and best mode of treatment is fixation of the thigh in extreme flexion against the abdomen by means of a wide strip of adhesive plaster passing from the back over the abdomen with the thigh bent upon it.

Occasionally a plaster-of-Paris dressing cannot be dispensed with for the transportation of such patients; of late it has found application for the purpose of enabling the patients to walk early and making the treatment ambulatory. Though these and similar (splint apparatus) methods are valuable, they are not as yet suitable for general medical practice.

When a fracture has united with marked displacement it must be again severed (osteoclasis or osteotomy) and the extension treatment carried out with exactness.

C. FRACTURES AT THE LOWER END OF THE FEMUR (PLATES 52, 53, FIG. 4; PLATE 57, FIG. 1).

These fractures are much rarer. Supracondylar transverse fractures may present a very pronounced displacement, which is effected in a typical manner by the influence of the calf muscles upon the lower fragment. The latter is thus put in flexion, and the two fragments override (Plate 52). A similar displacement has also been observed in traumatic separation of the epiphysis, though as a rule it is slighter, since the periosteal covering is partly preserved. Of course the traction of the thigh muscles tends to increase the displacement.

The examination demonstrates abnormal mobility at the lower end of the femur, especially characteristic in a transverse lateral direction; with this there is crepitation, which has a softer character in separation of the epiphysis. It is advisable to make the examination under anæsthesia.

In the treatment permanent extension is suitable, perhaps combined with gentle pressure (a wound roller) from behind against the lower fragment. At the same time it must not be forgotten that the above-named displacement of the lower fragment may cause very severe symptoms by pressure upon the large vessels or the sciatic nerve.

Fracture of one condyle is an intra-articular injury whose diagnosis is made by lateral to-and-fro movements at the knee, when crepitation and local pain may be elicited. As this fracture is very apt to be followed by varus or valgus position at the knee-joint, careful treatment is necessary, best by an extension dressing. If the effusion of blood in the knee-joint is profuse it should be evacuated by aspiration.

4. KNEE-JOINT.

A. LUXATIONS AT THE KNEE-JOINT.

Injuries to the ligamentous apparatus of the knee-joint are not so rare as true luxations of the articulation. What has been more frequently observed of late is displacement of the semilunar cartilages, especially after violent rotary movements with flexed knee. The external meniscus is more often affected than the internal. Conditional for the occurrence of these injuries is a certain relaxation of the ligaments, as well as at least a partial laceration of these bands. If the cartilage is truly luxated the knee is semi-flexed, fixed, and extension is impossible. Reduction under rotary movements after energetic distraction of the joint. Should the firmness of the articulation be impaired the cartilage is to be fixed in its normal position by operation, *i.e.*, by suture.

True luxations at the knee are very rare. The leg may be—

Luxated forward by hyperextension after laceration of the lateral and crucial ligaments;

Luxated backward, *i.e.*, this is rather a forward luxation of the condyles of the femur;

Luxated laterally, the leg being placed in abduction or adduction.

In all cases the condyles of the femur may be palpated more or less distinctly in their abnormal position. Owing to the peculiar force required to produce these luxations, it is inevitable that complicated injuries should be frequently present. Reduction is said to be easy, by traction and direct pressure.

B. LUXATIONS OF THE PATELLA (PLATE 56).

Dislocations of the patella are not the rarest of injuries. The fixation of the patella is not very strong; it is like that of a sesamoid bone interposed between the ligamentum patellæ and the quadriceps and only loosely fixed laterally.

Outward dislocation of the patella is the most frequent form; this is favored by its position, since it is always situated more over the external than the internal condyle, especially so if any valgus position is present at the knee. The luxation is incomplete when the articular surfaces are still partially in contact; complete, when the patella is dislocated entirely to the lateral surface of the external condyle. The injury may arise when the knee is extended and when it is flexed; in the former case the patella passes directly outward over the anterior surface of the lower margin of the femur (this may result from muscular action of the quadriceps, the knee being hyperextended); in the latter case the dislocation occurs in the groove between the external condyle and the tibia, not rarely through a force acting directly from in front inward. The diagnosis of this luxation is easy, since the patella is absent from its normal position and is felt in an abnormal location. Reduction by direct pressure, the knee being extended and the hip flexed, thus relaxing the quadriceps.

A vertical luxation is present when the patella is rotated 90° so that its edge lies in the depression between the two condyles of the femur. We distinguish an internal and an external vertical luxation, according to whether the cartilaginous surface of the patella is directed inward or outward. This injury results from a force acting directly from in front and laterally; it is said that it may be due also to pure muscular action. The position of the patella is easily recognized on the extended leg.

Complete inversion of the patella is a rotation by 180°, hence an increase of the afore-mentioned verti-

cal luxation. Then the articular surface of the patella points forward. The injury is extremely rare.
The diagnosis is difficult unless very exact palpation
is possible and the torsion of the quadriceps and of the
ligamentum patellæ can be recognized.

C. FRACTURES OF THE PATELLA (PLATES 55, 56, 57).

Fractures of the patella are not very frequent, but
they are of great interest. They result from direct
and from indirect force, and it is noteworthy that the
fracture of direct origin has usually a much better
prognosis than that occurring indirectly. A direct
lesion of the patella causes often a multiple, so-called
stellar fracture; indirect injury such as occurs mainly
through sudden muscular traction of the quadriceps
(fracture by traction), however, leads frequently to a
transverse fracture of the patella with a more or less
extensive laceration of the aponeurotic layers passing
alongside that bone. This circumstances in particular
is of great importance (Plate 55), since a marked gaping of the patellar fragments can result only when
these lateral layers are torn. Fracture by traction of
the patella occurs in a pronounced form by sudden
traction of the quadriceps in persons stumbling, etc.;
under such circumstances the tendon of the muscle
and that of the ligamentum patellæ resist the force;
in rare cases the tuberosity of the tibia is torn off,
more frequently the patella breaks.

The symptoms are very simple when the fracture,
as is usually the case, passes transversely through the
middle of the patella and is associated with some gap-

ing of the fragments. As the patella is fully inclosed in the joint capsule, such injury is a pure articular fracture; the effusion of blood is within the joint, and may exceptionally be quite considerable and fill the entire joint tensely. In recent cases it is possible as a rule to approximate the fragments so that they touch and produce distinct crepitation. When only a small marginal piece of the patella is separated, and in general when its periosteum is largely intact, the diagnosis may be difficult and doubtful.

Treatment.—In no other fracture is it possible to observe the fact common in this injury, that cases united with marked diastasis still functionate very well, while cases in which the fragments are well placed present occasionally serious and permanent impairment of the function of the limb. Very important in this connection is the condition of the quadriceps. This muscle presents in some cases the symptoms of pronounced atrophy due to prolonged inactivity and especially to reflex influences transmitted by the spinal centres. For this reason a mode of treatment has been developed in recent times, which, while dispensing with direct approximation of the fragments, directs its main efforts to the care of the quadriceps by massage (kneading and tapotement); this is done every day, and at the same time the fragments are moved toward each other, the leg being placed with knee extended and hip flexed, because in this position the quadriceps is relaxed. Though this method is correct and valuable, it must still be called rather one-sided, and there is no reason why it should not be combined with an attempt to approximate the fragments directly.

The unfavorable results following patellar fractures are unquestionably due to several causes. The traction of the quadriceps, with the diastasis of the fragments due to it, is an important factor, as is the atrophy of this muscle, which in some cases, even when the lesion of the patella is slight, may be extreme and perhaps irreparable. Of some importance, too, is occasionally the effusion of blood present in the joint, since it crowds the fragments apart. Unfavorable factors exist also in the fragments themselves, e.g., the scanty bone formation dependent on the fact that the patella possesses on one side a thick cartilaginous surface, on the other a (fibro-periosteal) fibrous layer. Then as a rule there is a kind of interposition of the fibres, elongated as they are by stretching until they have given way, of the external fibrous layer, which are situated over the fractured surfaces and imprisoned between them. It is this circumstance in particular which favors the occurrence of a ligamentous union even when the fragments are in good apposition.

The treatment of course must strive to overcome these obstacles as much as possible. The leg is to be fully extended at the knee and bent at the hip so as to relax the quadriceps. The knee-joint is fixed by a posterior splint, e.g., of plastic felt which is moulded warm. The fragments are approximated manually and freed by lateral friction as well as may be of the interposed tissues, and then kept in position by strips of adhesive plaster applied in the form of slings and crossing on the posterior surface over the splint. If the effusion of blood is great it is removed by aspiration. The quadriceps is masséed daily by kneading and tapotement, mainly in a downward direction so as to depress the upper fragment.

A failure of any kind of union between the fragments and an adhesion of the upper fragment to the anterior surface of the thigh are rare occurrences, both equally unfavorable.

In severe cases operative interference cannot be dispensed with. The fragments may be united subcutaneously by a tendon suture. Some prefer the use of the old clamp of Malgaigne. Direct bone suture of the fragments of course is the most reliable procedure, but it should be performed only by a skilled surgeon.

5. LEG.

A. Fracture of Both Bones in the Region of the Diaphysis (Plates 58, 60).

This injury is very frequent. It results mainly from a direct force (run-over accidents, etc.), and affects both bones at about the same point. Indirectly, especially by torsion of the body while the foot is fixed, isolated oblique fractures in the lower part of the tibia often result, and the fibula then breaks only in consequence of the weight of the body which it is too weak to bear; the fracture then is generally incomplete and frequently is higher up on the shaft. Of course oblique fractures (by flexion or torsion) are in general somewhat more unfavorable than transverse fractures, and the fragments are more liable to be displaced. Frequently the pointed upper fragment, double-pointed when the line of fracture in front terminates exactly at the crest of the tibia (in the form of the mouthpiece of a clarionet), presses forward against the skin and may perforate it.

The diagnosis of the fracture is usually very easy, because abnormal mobility, crepitation, and displacement are readily demonstrated. Torsion of the lower fragment, if present, will be recognized by noting the position of the patella and of the foot, and by careful palpation of the crest of the tibia from above and below as far as the point of fracture. To locate the point of fracture at the fibula sometimes requires considerable skill.

Treatment.—The most exact reposition possible should always be attempted by vigorous traction upon the injured foot, counter-extension at the thigh or pelvis, and direct manipulation at the seat of the fracture. When the fracture is oblique the displacement is very liable to recur. Protrusion of the upper, or exceptionally of the lower, fragment forward against the thin skin is to be met by appropriate position in slight hyperextension at the seat of the fracture. During the first week the most suitable is the so-called T-splint of Volkmann, made of stout tin; of course it is to be so well padded that no injurious pressure is exerted at any point, particularly the region of the heel.

In most cases of this kind a careful examination and reposition under anæsthesia at the end of the first week cannot be dispensed with, in my opinion. Then a padded plaster-of-Paris dressing is very useful; a second inspection must be made about a week later. Lateral deviations are easily prevented in this way; any rotation present requires even greater care. The occurrence of a hyperextension at the point of fracture is to be particularly watched for, otherwise a recurved position may remain behind.

For repressing the point of a fragment a special auxiliary should be mentioned, namely, Malgaigne's screw, which is fixed in the dressing and made to exert direct pressure on the protruding fragment by means of a movable stylus. Good reposition, appropriate position, and in some case the application of permanent extension by weights will suffice as a rule.

After the fracture has united, which process is favored or hastened by an ambulatory dressing, functional restoration should be promoted by baths, massage, active and passive exercises of the joints. Should a disturbing or painful bony prominence have remained at the site of the fracture it had best be removed by the chisel; the place should be laid bare by a flap incision.

B. Isolated Fracture of the Tibia (Plates 57, 59, 60).

a. Fracture of the tibia at its upper end (Plate 57, Fig. 3) is usually the result of compression, *i.e.*, the articular end of the tibia suffers an infraction by the pressure of the opposite condyle of the femur. This may be due to a fall upon the foot from a considerable height. Once I observed this fracture as a result of springing from a bicycle. The symptoms presented are those of a severe distorsion or contusion of the joint. Movements in the joint are painful; lateral to-and-fro movements are usually possible, and at the upper end of the tibia distinct painful points are found. The fracture generally affects only one half of the articular surface, and hence is very apt to cause a varus or valgus position in the knee-joint. After infraction of the inner half of the upper articular surface of the tibia a varus position will easily result and remain behind unless specially prevented. Treatment by permanent extension by weights and pulleys, with a sliding foot-board, perhaps combined with lateral traction by a sling which over-corrects the threatened anomalous position. Of course there should be early massage and mobilization, as in all articular fractures.

b. Traumatic separation of the epiphysis at the upper end of the tibia is a very rare injury. The symptoms present point to the possibility of this lesion in a case of marked contusion at the upper end of the tibia in a child. A positive diagnosis is possible only under anæsthesia, when abnormal mobility and characteristic cartilaginous crepitation can be demonstrated. Treatment on general principles.

c. Separation of the tuberosity of the tibia, a very rare injury in children (in the form of a separation of an epiphysis or apophysis) and in adults. The fragment is drawn upward by the traction of the quadriceps; active extension of the leg in the knee-joint is impossible. The fragment is felt under the skin and is easily movable in all directions. The patella can be felt above to be intact. The knee-joint need not necessarily be implicated, but usually contains an effusion of blood. The treatment may be similar to that of fractures of the patella; nailing the accurately replaced fragment to the tibia is the best procedure.

d. Fracture of the Shaft of the Tibia.

It has been stated above that in fracture of both leg bones not rarely the tibia breaks first, fracture of the fibula resulting secondarily. Aside from torsion this may also occur by flexion; for we may observe often enough in osteoclasis of rachitic leg bones that the tibia alone breaks, and that additional force is necessary to fracture the fibula likewise. The symptoms as a rule are clear and easily comprehended. When isolated fracture of the tibia is associated with marked displacement, the fibula, which would serve as a sort of splint, must be implicated. The fibula then must either be also completely broken and present a similar displacement, or—and this occurs chiefly in fracture of the tibia in the upper half of the shaft—it is luxated. Thus the capitulum of the fibula is found dislocated upward (Plate 59). In the more recent cases this can be overcome by careful reposition; for retaining the fragments a well-fitting plaster-of-Paris dressing or permanent extension by weights is useful.

C. ISOLATED FRACTURE OF THE FIBULA.

A very rare lesion which can result only from violent direct force. The capitulum of the fibula may be torn off by violent traction of the biceps femoris. Occasionally the peroneal nerve is injured. There is little tendency to displacement. Treatment on general principles.

D. FRACTURES AT THE LOWER END OF BOTH BONES.

a. *Typical Fracture of the Ankle* (*Plates* 61, 62, 63).

This fracture may be compared to the typical fracture of the epiphysis of the radius; as in the latter, the mode of occurrence, the symptoms, and the principles of treatment have a typical character. That in fracture of the ankle the fibula breaks likewise is readily understood by reason of the anatomical arrangement, *i.e.*, on account of the firm union of tibia and fibula at their lower end.

Typical malleolar fracture results from outward flexion of the body when the foot is fixed or from flexion of the foot outward. In the latter way these fractures may be produced in the cadaver: the leg is so placed as to rest on the outer surface, the foot with the malleolar region projecting over the edge of the table; a vigorous push brings the weight of the experimenter's body to bear on the foot, which assumes a certain abducted position, the internal malleolus breaks off, and, the force continuing to act, the fibula fractures slightly above the external malleolus, corresponding to the edge of the table.

In the majority of malleolar fractures we find the conditions exactly similar. The movement of abduction of the foot in the astragalo-crural joint causes great tension of the internal lateral or deltoid ligament; if the movement continues, as a rule the ligament is not lacerated, but the tip of the malleolus is torn off. Now the force continuing acts on the foot

as a whole and especially crowds the astragalus against
the external malleolus and produces the fracture above
the latter by flexion. In some cases, too, the weight
of the body after separation of the internal malleolus,
the foot being abducted, produces fracture of the fibula
by flexion, as the latter bone alone is too weak to bear
the weight.

Symptoms.—In typical fracture of the ankle, therefore, we find the tip of the internal malleolus abnormally movable and often displaced downward, while
the fibula is broken above the external malleolus. If
the examiner taken the foot in his hand and at the
same time fixes the leg above the malleolar region, he
can produce an abnormal lateral displacement, especially an abduction (pronation) of the foot to an un-
usual degree. The foot, moreover, by itself occupies
an abnormal position, a kind of valgus position, an
outward deviation. The region of the internal malleolus, or more correctly the fractured edge of the tibia,
sometimes projects so markedly that the thin overlying integument is very tense and threatens to give
way; if it is torn, *i.e.*, if a compound injury has resulted, an actual luxation is not rarely present. The
lower end of the tibia may project through the integument so far that reposition can be effected only
after extensive division of the interposed skin. On
the fibula the characteristic infraction above the
malleolus is always more or less pronounced.

It is very important to picture to one's self accurately the anatomical details of this fracture. The
piece torn off the internal malleolus is sometimes very
small. The breaking off of the fibula in the manner
described is possible, of course, only by a solution of

the firm ligamentous connection between tibia and
fibula at their lower end. These ligaments may tear;
but a larger or smaller piece of bone may be torn off
from the articular end of the tibia at the same time.
In this way fragments are separated in front by the
anterior tibio-fibular ligament and sometimes also
behind by the posterior ligament of the same name.
(See Plate 62.) It is only after the connection between tibia and fibula has been severed that the latter
can be bent sidewise so as to produce a complete or
incomplete fracture of this bone.

Prognosis.—The typical malleolar fracture, even
if not complicated, is always a severe injury. It is a
true articular fracture, and doubly important for the
fact that the affected joint has to bear the weight of
the entire body. In the treatment, even at the present
day, serious mistakes are sometimes committed which
jeopardize the function of the joint and the working
capacity of the patient.

Treatment.—The first requirement is an exact reposition of the fragments. The foot as a whole must
be displaced toward the tibia in the sense of an adduction. Formerly stress was laid on the fact that
the foot should also be brought into a true varus position so as to overcome or prevent the present or im-
pending valgus position. This is not necessary if the
foot itself is exactly replaced, whereby of course the
angle of infraction at the fibula above the external
malleolus must be completely effaced. Sometimes it
is still more important to counteract a backward displacement of the foot, which is likewise present, by
forward traction.

After reposition, if necessary under anæsthesia, the

foot and leg must be placed at rest, for which purpose in the first few days a Volkmann tin splint and later a Beely's plaster-of-Paris splint are most appropriate. During the first two weeks the dressing should be removed every three or four days, later every other day, with a view of instituting massage of the joint and passive movements; during these manipulations the position of the foot should receive attention, for I know of cases in which a good position of the foot, which was present in the early weeks, became impaired subsequently by reason of lack of care in the dressing. This point should receive attention even later on. After the fracture is consolidated and the patient attempts to walk, the need of a protective splint is still present, and he should not be discharged without a fitting shoe in order to prevent the occurrence of pes valgus. Of late I have observed good results from the application of medico-mechanical apparatus.

When we have to deal with an unfavorable position of this fracture, and the fragments have been fixed for weeks in the objectionable position, appropriate operative treatment must be at once resorted to. If the new connections can no longer be broken by simple fracture, osteotomy of the fibula at the site of the fracture and sometimes also of the internal malleolus should be performed so as to replace the foot.

When the foot shows a persistent tendency to assume a valgus position, Dupuytren's old splint dressing will still be found useful. This consists in the application to the inner side of the leg of a splint which is so padded and fixed as to extend beyond the

region of the internal malleolus and the foot, so that
the latter may be drawn by turns of a bandage toward
the splint. It is clear that in this way a powerful
obstacle is opposed to an outward displacement and a
valgus position of the foot.

b. Other malleolar fractures result from adduction or supination of the foot, whereby the tip of the
external malleolus is first torn off and a varus position
is apt to occur; or else, from torsion of the foot in the
astragalo-crural joint, whereby fractures by torsion
and flexion of the tibia may occur, as well as fracture
of the fibula. These injuries cause no great difficulties when the examination is careful, and should be
treated in a manner similar to typical malleolar fracture. The same remark applies when only one malleolus is fractured; the acting force having been less
intense or the injury having resulted from indirect
influences.

c. Separation of the Epiphyses at the Lower End of the Leg Bones (Plate 63).

This is a rare injury which of course occurs only
in children. Sometimes it is observed after forcible
redressement of severe clubfoot. It is diagnosticated
by the presence of abnormal mobility above the ankle
region, together with cartilaginous crepitation. The
treatment requires rest, followed by exercise.

d. Supramalleolar Fracture of Both Leg Bones (Plate 60, Fig. 3).

In this fracture, which is not apt to cause difficulty
in the way of diagnosis, we find a similar tendency
to displacement as in typical malleolar fractures.

The foot with the lower fragments is very liable to be displaced backward, which is to be especially prevented. Careful reposition and treatment on the principles explained under the head of typical malleolar fractures.

6. ANKLE-JOINT.

It is well known that the movements of the foot in the way of flexion and extension are effected in the astragalo-crural joint; those of pronation and supination in the astragalo-tarsal joint. In the latter case the movement is of such nature that the astragalus remains firmly connected with the leg bones; the articulations involved being those between the astragalus with the os calcis and the navicular bone. Excessive movements cause distorsion in the joints and eventually luxations.

a. Luxations in the Astragalo-Crural Joint (Plate 64).

These are the true luxations of the foot. They may be forward (by excessive dorsal flexion) and backward (by excessive plantar flexion). The position of the foot is so characteristic (see Plate 64) that the diagnosis is made without difficulty. Reduction is effected by direct pressure upon the tibia forward or backward, with simultaneous flexion in the direction which caused the luxation. Fracture of one malleolus during this manipulation is of no great consequence. Lateral luxations are not possible without malleolar fractures.

b. Luxation in the antragalo-tarsal joint, the

so-called luxatio sub talo, may be outward from excessive pronation, or inward from forced supination
of the foot. Much rarer are forward and backward
luxations in this articulation. The diagnosis may be
quite difficult; exact palpation of the bony prominences, the demonstration of abnormal mobility in the
astragalo-crural joint, observation of the altered form
of the foot, and especially examination under anæsthesia will settle it. Reposition is difficult; it requires complete relaxation of the muscles and the use
of appropriate movements aided by direct pressure.

c. Isolated Luxation of the Astragalus.

The astragalus may be dislocated in various directions. The mechanism is certainly a very complicated
one and thus far not fully elucidated. Marked deformity is present; the astragalus can be felt more or
less distinctly through the soft parts. The tibia is
approximated to the sole of the foot and sometimes
articulates directly with the os calcis.

Reduction is difficult. If it fails, it must be forced
by means of an incision, as in the luxations named
above. It is a noteworthy fact that good results are
obtained by aseptic treatment, although the astragalus
may have lost some of its connections and paths of
nutrition.

7. THE FOOT.

A. Fracture of the Tarsal Bones.

a. Fracture of the Astragalus.

During forced movements such as lead to luxation
at the tarsus we also meet with infractions, sepa-

ration of fragments, and fractures of the astragalus. Compression and fracture of the astragalus without associated luxation sometimes results from a fall upon the foot. The symptoms are those of a severe distorsion, and the diagnosis therefore remains uncertain. Treatment on general principles.

b. Fracture of the Calcaneus.

A fracture by traction occurs on the tuberosity of the os calcis, which may be wrenched off by sudden action of the calf muscles. The fragment is drawn up by the muscles. It can be replaced when the knee is flexed, and may be fixed by nailing.

Fracture by compression of the calcaneus results from a fall upon the feet. The bone is comminuted by the astragalus, which, acting as a wedge, forces the calcaneus apart. The fragments, which are usually numerous, in their displaced position cause a widening of the bone below the normal malleolar region. In this way a traumatic flat-foot may result, which may be slow to heal.

Fracture of the sustentaculum is very rare and cannot be positively diagnosticated; it results from forced adduction of the foot and produces a valgus position, with local pain.

c. Fracture of the Remaining Bones.

Fracture of the remaining tarsal bones is exceedingly rare; that of the metatarsals and the phalanges is of no practical importance, is generally easily diagnosticated, and its treatment is simple.

B. LUXATIONS.

a. *Luxation of the Tarsal Bones*

is a very rare injury. The diagnosis is based on the palpation of the dislocated bone. Reduction by direct pressure, if necessary through an incision.

b. *Luxation of the Metatarsal Bones,*

i.e., in Lisfranc's articulation so called, is met with mainly in the form in which the metatarsals are dislocated upon the dorsum of the foot. Reduction is difficult; each bone may have to be replaced separately.

c. *Luxation of the Toes.*

These injuries are analogous to those of the fingers, and of course are much rarer than the latter. Forced dorsal flexion causes an upward dislocation of the phalanges. The diagnosis is easy, and reduction is effected by pushing the dorsally flexed phalanges forward.